WARRIOR'S DISHONOUR

For Dimitris Georgellis and Dimitris Afendoulis

Warrior's Dishonour
Barbarity, Morality and Torture in Modern Warfare

Edited by

GEORGE KASSIMERIS

ASHGATE

Published by
Ashgate Publishing Limited
Gower House
Croft Road
Aldershot
Hampshire GU11 3HR
England

Ashgate Publishing Company
Suite 420
101 Cherry Street
Burlington, VT 05401-4405
USA

Ashgate website: http://www.ashgate.com

British Library Cataloguing in Publication Data
Warrior's dishonour : barbarity, morality and torture in
    modern warfare
    1. War - Psychological aspects 2. War - Moral and ethical
    aspects 3. Atrocities 4. Torture 5. Control (Psychology)
    I. Kassimeris, George
    355'.0019

Library of Congress Cataloging-in-Publication Data
Warrior's dishonour : barbarity, morality and torture in modern warfare / edited by
George Kassimeris.
        p. cm.
    Includes index.
    ISBN-13: 978-0-7546-4799-7
    ISBN-10: 0-7546-4799-4
    1. War--Psychological aspects. 2. Torture. 3. Human rights. 4. Violence--Case
studies. I. Kassimeris, George. II. Title: Warrior's dishonor.

    U22.3.W35 2006
    303.6'6--dc22

                                                                    2006021139

ISBN-13: 978-0-7546-4799-7
ISBN-10: 0 7546 4799 4

Printed and bound in Great Britain by MPG Books Ltd, Bodmin, Cornwall.

Contents

List of Contributors

Huw C. Bennett is a doctoral candidate in the Department of International Politics, University of Wales, Aberystwyth, and was recently an ESRC Visiting Scholar at the Centre for Military and Strategic Studies, University of Calgary. His thesis examines the conduct of the British Army towards non-combatants during the Kenya Emergency, and the various forces which led to atrocities there.

Ian Germani is Associate Professor in the Department of History at the University of Regina, Saskatchewan, Canada and has been Department Head since 2004. He is a specialist in the history of the French Revolution. His recent publications have focused on the relationship between war and culture during the Revolution. He is also working on a study of military justice in the armies of the French Revolution.

Dimitrios Giannoulopoulos is a Lecturer in Law at Brunel University in London, where he teaches Criminal Law, Evidence and International Human Rights. He possesses an LL.B. and an LL.M. in Criminal Law and Procedure from the University of Athens. He also holds an LL.M. (D.E.A.) in Criminal Studies and Criminology as well as a Diploma of Comparative Legal Studies from the University of Aix-Marseille III. He received an M.Phil in Evidence from Brunel University in 2002. Mr Giannoulopoulos is now concluding his doctoral thesis at the *École Doctorale de Droit Comparé*, at the University of Panthéon-Sorbonne (Paris I). His research focuses on the comparison between automatic and discretionary models of exclusion of unlawfully obtained criminal evidence. It draws on an in-depth analysis of the models adopted by England, France, Greece and the United States.

Richard Jackson is Lecturer in International Security at the University of Manchester. His research focuses on critical terrorism studies and the causes of political violence. His most recent book is *Writing the War on Terrorism: Language, Politics and Counter-terrorism* (Manchester University Press, 2005). He is currently preparing a monograph for Manchester University Press entitled, *What Causes Intrastate War? Towards an Understanding of Organised Civil Violence*.

Graham Long is a British Academy Postdoctoral Research Fellow in the School of Geography, Politics and Sociology at Newcastle University. He is the author of *Relativism and the Foundations of Liberalism* (Imprint Academic, 2005). His current research interests focus on the relationship between relativism and questions of international justice, particularly cosmopolitanism and just war theory.

Tim Montgomery's varied research interests include the use of political violence, the role of non-state agents, international human rights and humanitarian standards, forced migration and forms of global power. Montgomery holds two Masters, the first in Refugee Studies, the second in International Relations Research. He is presently completing a PhD at the University of Sheffield, UK, observing what armed resistance may tell us about the nature of globalizing power. He has worked for many years in NGOs, in the UK and Sierra Leone, focusing on peace, human rights and refugees.

Michael Plaxton is a Lecturer in Law at the University of Aberdeen, teaching jurisprudence and various subjects associated with criminal justice. He received a Doctorate in Juridical Science from the University of Toronto in 2004. He has published articles pertaining to legal theory, constitutional law, criminal law and evidence in a number of Commonwealth and American journals.

Stephen Riley is Lecturer in Philosophy at St. Martin's College. He received his doctorate – a phenomenology of international criminal justice – from Lancaster University. His current research concerns the application of critical theory to international law.

Uwe Steinhoff is Research Associate in the *Oxford Leverhulme Programme on the Changing Character of War* and Affiliated Researcher at the *Oxford Uehiro Centre for Practical Ethics*. He is the author of the books *Kritik der kommunikativen Rationalität* (on Habermas and Apel) and *Moralisch korrektes Töten: Zur Ethik des Krieges und des Terrorismus* (*Morally Correct Killing: On the Ethics of War and Terrorism*, forthcoming) as well as of articles on epistemology, postmodernism, ethics and political philosophy. He is currently working on a book on war and non-state actors and on issues of global justice.

Frank Tallett is Senior Lecturer in Early Modern European History at the University of Reading and co-Director of its Centre for the Advanced Study of French History. His twin research interests are in early-modern warfare and the history of French Catholicism under the old regime. His books include *War and Society in Early Modern Europe, 1495-1715* (London and New York, 1992/2001), *Priests, Prelates and People: A History of European Catholicism since 1750* (co-author, London and New York, 2003), *The Right in France from the Revolution to Le Pen* (co-editor, London and New York, 2003), *Catholicism in Britain and France since* 1789 (co-editor, London and Rio Grande, 1996), and *Religion, Society and Politics in France since 1789* (co-editor, London and Rio Grande, 1991).

Anthony Vinci is a PhD candidate in the International Relations department at the London School of Economics, where his research focuses on the international relations of armed groups. He recently returned from field work on the Lord's Resistance Army (LRA) in northern Uganda. Before commencing his PhD, Anthony was a consultant at Toffler Associates, the strategic consulting firm founded by Alvin and Heidi Toffler.

David Whetham initially took a degree in Philosophy at the London School of Economics before going on to take a Masters Degree in War Studies at King's College London (KCL). After some time spent travelling around the Great Lakes region of Africa, David returned to KCL to take a PhD in War Studies. While studying, David worked as a tutor and lecturer, spent a year as a researcher for BBC History and was employed by the OSCE in Kosovo, contributing to the 2001 and 2002 elections. In 2003 David joined KCL's Defence Studies Department, based at the Joint Services Command and Staff College, where he concentrates on the ethical, legal and moral dimensions of war.

Acknowledgements

In a collaborative volume, it is for individual contributors to express their gratitude in their chapters. The editor, however, should still fully exercise his or her right to a list of acknowledgements in a larger typeface and I would like to take this opportunity to thank Carol Millwood, my best friend, mentor and most forthright critic, who took time out of her busy schedule to read parts of the manuscript and correct a host of errors at extremely short notice. I would also like to thank my editors at Ashgate, Kirstin Howgate, who enthusiastically supported this project from the outset, and Sarah Cooke, who guided the manuscript to publication with supreme patience and skill. Working with Sarah has been a pleasure. I am also grateful to the British Academy for their generous assistance in funding this project.

My thanks goes also to all my friends and colleagues at Wolverhampton for their help and advice, especially to Malcolm Wanklyn, Mark Phythian, Yolanda Pascual-Sole, Dieter Steinert, John Benson, Deirdre Burke, Eammon O'Kane, Penny Welch, John Buckley, Tony Shannon-Little, Janine Tellet, Tracy Begum, Jon Moran, William Pallet, Martin Durham, Jean Gilkison, Barbara Gwinnett, Emma Kilvert, Maggie Russ. Away from Wolverhampton, special thanks are due to Conor Gearty, Joanna Bourke, Richard Overy, Paul Rogers, Hew Strachan, Marilyn Young, Frank Faulkner and Maria Angelaki.

My deepest thanks, however, go to Kuzi Kassimeris-Salinas, love of my life, for her support and confidence in me over the years and for tolerating my sometimes unreasonable working regime and irascibility that comes with it.

Wolverhampton, September 2006   G.K.

Chapter 1

The Warrior's Dishonour

George Kassimeris

'What does the earth look like in the places where people commit atrocities?' wondered American writer Robert Kaplan while researching his book on the Balkans and its people. 'Is there a bad smell,' he asked, 'a genius loci, something about the landscape that might incriminate?'[1] It is probably tempting to think that yes, in places whose names have become synonymous with the atrocities of our times and where hundreds of thousands have lost their lives, a permanent ghastly darkness coupled with a dull smell of damp and rot does exist. This is not true, of course. Take Rwanda, for example. Rwanda, in Philip Gourevitch's reconstruction of the 1994 events, is described as spectacular to behold.[2] 'Throughout its center,' he writes, 'a winding succession of steep, tightly terraced slopes radiates out from small roadside settlements and solitary compounds. Gashes of red clay and black loam mark fresh hoe work; eucalyptus trees flash silver against brilliant green tea plantations; banana trees everywhere.'[3] A beautiful country by all accounts but also a country which went through the horrific trauma of neighbour killing neighbour. All told, an estimated 800,000 Rwandans were killed – slaughtered, may be a more accurate phrase – in fewer than a hundred days. What happened in Rwanda was a shameful passage in twentieth century history but it was not an isolated incident of aberrant behaviour. Bosnia, Sierra Leone, Somalia, East Timor, Darfur, Kenya, Algeria, Cambodia, Chechnya – the list could go on and on – have all witnessed indiscriminate waves of killing of the most horrifying kind.

What drives ordinary people into hatred, genocide, inhumanity and evil? What turns friends and neighbours against each other with such savagery? What turns fresh-faced boys into killers of people who have done them no harm? Where does such barbarity come from? This collection of essays is about the anarchy, cruelty and overwhelming confusion of modern warfare. It is also about man's savagery and inhumanity to man in times of war. The characteristic act of men at war, to paraphrase Joanna Bourke's opening sentence in her magisterial *An Intimate History of Killing*, is not killing: it is killing by committing shocking and unspeakable atrocities, when

---

1   Robert D. Kaplan, *Balkan Ghosts: A Journey Through History* (London, 1993).

2   Philip Gourevitch, *We Wish to Inform You that Tomorrow We will be Killed with Our Families* (London, 1999).

3   Ibid., p. 20.

circumstances permit.[4] If there were any doubts about mankind's capacity for sheer bestial savagery, the twentieth century's excessive violence, barbarism and genocide put them to rest. From the bloodshed of the Western Front to the massacres of the Armenians, from Stalin's camps to the rape of Nanking, from the butchery of Bosnia to the slow motion genocide of Darfur, men and women of all creeds and colours exhibited a staggering appetite for death and destruction. That said, barbarity in warfare is hardly an exclusively twentieth century phenomenon. True, the last century will go down in history as one of the most gruesome and murderous centuries but the exercise of indiscriminate terror, ethnic cleansing, genocide and rape as war-making tools has been used for millennia.

The romance of warfare, however, should never be underestimated. 'Three thousand years,' the classicist Bernard Knox once commented, 'have not changed the human condition: we are still lovers of the will to violence.'[5] Why is war so seductive? One of the many reasons why *Apocalypse Now* is remembered when dozens of other war movies, from *The Boys in Company C* to *Go Tell the Spartans*, are pretty much forgotten is that it embraces the full emotional range of the phenomenon and dares to admit something that less ambiguous works cannot allow themselves to countenance: seen from a certain angle, war can be intoxicating and exciting. Colonel Kilgore's dawn helicopter raid and his notorious line about remembering the smell of napalm – 'smelled ... like Victory' – says a great deal more about the emotional texture of war and the condition of men in war than the mesmeric appeal of technologies of killing.

How people explain and justify war does not necessarily account for why they wage it. Despite the uncertainty, the fear of death and the catastrophe of defeat, warfare has always attracted people. No matter how many times the nature of the argument about the use and value of warfare has changed over the centuries, war fascinates men more than it repels them. When Alexander the Great took a copy of the Iliad with him on campaign, it was not because it served as a cautionary reminder of the bitterness and folly of war. Homer describes the grief and lamentation of mothers and fathers, comrades and lovers; but the Iliad as a whole celebrates heroism rather than horror and violence.

How does one explain why war 'no matter how futile, repulsive, or dysfunctional'[6] it may be, remains, to use the words of Michael Ignattief, one of the 'ecstatic activities of mankind like sport and sex.'[7] This seems an obvious question to ask; but in many ways it is more complex than it looks. To be precise, the question should be phrased differently: why is it that men cannot live without war? There have been attempts to

---

4   Joanna Bourke, *An Intimate History of Killing: Face-to-Face Killing in Twentieth Century Warfare* (London, 1999).

5   Bernard Knox, *Essays Ancient and Modern* (Baltimore, 1990), p. 98.

6   Barbara Ehrenreich, *Blood Rites: Origins and History of the Passions of War* (New York, 1997), p. 18.

7   See Michael Ignattief's review essay 'The Gods of War' in *New York Review of Books*, 9 October 1997.

answer the question primarily by turning to the study of human culture to identify the 'moment' when the urge to war became part of human nature but such an approach fails to go even near to explaining people's strange attraction to warfare.[8] War's enduring attraction, writes *New York Times* veteran foreign correspondent Chris Hedges in his powerful chronicle of modern war, is that it gives people's lives meaning:

> Even with its destruction and carnage it can give us what we long for in life. It can give us purpose, meaning, a reason for living. Only when we are in the midst of conflict does the shallowness and vapidness of much of our lives become apparent … War makes the world understandable, a black and white tableau of them and us.[9]

People, as Vietnam War veteran William Broyles Jr so eloquently put it, 'love war in strange and troubling ways.'[10] In Broyles' view, it is men in particular who love war because they love games: 'war is a brutal, deadly game, but [still] a game, the best there is.' 'At some terrible level, war is for men,' says the Vietnam veteran, 'the closest thing to what childbirth is for women: the initiation to power of life and death. It is like lifting off a corner of the universe and looking at what's underneath.' But beyond the intense experience of battle and the heightened excitement that apparently goes with it, war, the Vietnam veteran says, 'is an escape from the duties of everyday life, from the bonds of the family, community and work.'

Matt Pottinger, a *Wall Street Journal* journalist, is a strong case in point. Pottinger left the newspaper in 2005 to join the US Marines. In his last piece for the paper entitled 'Mightier than the pen', Pottinger explains his career change:

> Friends ask me, if I worry about going from a life of independent thought and action to a life of hierarchy and teamwork. At the moment, I find that appealing because it means being part of something bigger than I am. As for how different it's going to be, that too, has its appeal because it's the opposite of what I've been doing up to now. Why should I do something that's a 'natural fit' with what I already do? Why shouldn't I try to expand myself? In a way, I see the Marines as a microcosm of America at its best. Their focus isn't on weapons and tactics, but on leadership. That's the whole point of the Marines. They care about each other in good times and bad, they've always had to fight for their existence …[11]

Yet, the essence of war remains the organized maiming, crippling and murder of other human beings and that is a point well made by another marine who is also good

---

8 Apart from Ehrenreich's, *Blood Rites*, see also Philippe Delmas, *The Rosy Future of War* (New York, 1997); Vannevar Bush, 'Can Men Live Without War?', *Atlantic Monthly* (February 1956); see Niall Ferguson's review of *An Intimate History of Killing* entitled 'Close Encounters of a Lethal Kind' in *The Spectator*, 10 July 1999.

9 See Chris Hedges, *War is a Force that Gives us Meaning* (New York: Public Affairs, 2002), p. 14.

10 William Broyles, 'Why Men Love War' in *Esquire* magazine (1984) and in Walter Capps (ed.), *The Vietnam Reader* (New York, 1991).

11 *Wall Street Journal Europe*, 15 December 2005.

with words, Anthony Swofford, in his book *Jarhead: A Soldier's Story of Modern War*, a powerful memoir which combines the black humour of *Catch-22* with the savagery of *Full Metal Jacket* and the visceral detail of *The Things They Carried*. Swofford joined the US Marine Corps at the age of 17 and the first thing he learned was that 'to be a marine, a true marine, you must kill … if you don't kill, you're not a combatant.' Two months after his twentieth birthday, he was stationed in Riyadh awaiting the onset of the 1991 Desert Storm, in what we now call the first Gulf War. A decade after that, and armed with a degree from Iowa's Writers Workshop, Swofford wrote *Jarhead* in which he reaches some unpleasant conclusions about war and warriors.

> I have gone to war and now I can issue my complaint. I can sit on my porch and complain all day. And you must listen. Some of you will say to me: You signed the contract, you crying bitch, and you fought in a war because of your signature, no one held a gun to your head. This is true, but because I signed the contract and fulfilled my obligation to fight one of America's wars, I am entitled to speak, to say, *I belonged to a fucked situation*. I am entitled to despair over the likelihood of further atrocities.[12]

Swofford does, it scarcely needs saying, make the point that some wars are unavoidable and need to be fought, but at the same time he makes us see there are no limits to which men could not be driven once the chaos of warfare opens up Pandora's box. Which begs the most crucial question of his book 'will the wars of the world ever end?' The marine turned author pulls no punches in his answer: 'Sorry, we must say to the mothers whose sons will die horribly, this will never end.'[13]

'The Enemy is not a Human Being'

All war cheapens human life, all brutalizes and removes the taboo of violence. But can warfare be anything else than barbaric? Much historical evidence, some of it to be drawn from the pages of this collection, shows that there is not much in recent history to prove that it can. The kind of dark barbarity that defined much of the world before the creation of the nation-state, has to large degree characterized the world that came after. The remarkable thing, however, is that it was only towards the end of the twentieth century that people in the West began to understand a basic fact that Sri Lankans, Haitians, Liberians, Afghans, Chechnyans, Cambodians, Angolans and many others have long known all too well: that warfare prosecuted according to recognized laws of war has been the exception not the rule.

For centuries we have debated the morality of going to war and the manner in which it is fought, but international conventions are not sufficient in themselves to

---

12 Anthony Swofford, *Jarhead: A Soldier's Story of Modern War* (London, 2003), p. 254.

13 Ibid., p. 255.

make warriors adhere to the rules.[14] People have continued to commit war atrocities, refusing to distinguish between combatants and non-combatants, legitimate and illegitimate targets and civilized and barbarous treatment of prisoners and of the wounded.

In the light of this it might be not be difficult to explain why the ratio of military to civilian casualties since the First World War has risen so dramatically that civilian casualties now constitute the majority – an astonishing and depressing 90 per cent – of those killed, mutilated, raped and uprooted even when they presented no conceivable threat to the military adversaries. What is even more depressing is the fact that a large number of them are children. Yet, what is less well-known is that a growing number of the people behind the killings are children themselves. As P.W. Singer points out in his fascinating, informative but equally disturbing study *Children at War*, child soldiers, some of whom are no older than six, are to be found in three quarters of the world's current 50 or so conflicts.[15] In Sierra Leone's Revolutionary United Front (RUF), 80 per cent of the fighters were aged between seven and 14.[16] In Liberia, 20,000 children are reported to have served in the country's protracted civil wars. Recruited from orphanages, refugee camps, the slums of impoverished cities, or among those made destitute by AIDS, famine and war, children are easy to brainwash, quick to train, and readily drugged, terrorized and conditioned into committing reckless atrocities.[17]

What can be done once the point has been reached when young boys and girl soldiers carrying Kalashnikovs larger than themselves viciously attack villages full of old people and mothers with their newly-born children? The answer, of course, should be a great deal if only we were able to explain both the complexity and psychology of such ugly brutality. Yet, explaining evil has turned out to be as difficult as preventing it. For if we truly understood the causes of barbarity, human behaviour would have improved over the ages alongside our technology.

Nobody has yet come up with a coherent plan for understanding and dealing with the murky depths of the human psyche and Joanna Bourke is probably right in arguing that 'it is not necessary to look for extraordinary personality traits or

---

14 See Geoffrey Best's classic *Humanity in Warfare: The Modern History of the International Law of Armed Conflicts* (London, 1980); for a more contemporary take on the subject see Rupert Smith's *The Utility of Force: The Art of War in the Modern World* (London, 2005), pp. 377-383; The best available treatment of the history of laws of war, however, is a collection edited by Michael Howard, George Andreopoulos and Mark R. Shulman, *The Laws of War: Constraints on Warfare in the Western World* (London, 1997).

15 P.W. Singer, *Children at War* (New York, 2005).

16 See Alephonsion Deng, Benson Deng and Benjamin Ajak, *They Poured Fire on us from the Sky: The True Story of Three Lost Boys from Sudan* (New York, 2005); Richard Maclure and Myriam Denov, '"I Didn't Want to Die So I Joined Them": Structuration and the Process of Becoming Boy Soldiers in Sierra Leone' in *Terrorism and Political Violence* (Vol.18, No.1, 2006), pp. 119-136.

17 See *The Scars of Death: Children Abducted by the Lord's Resistance Army in Uganda*, a report by Human Rights Watch, September 1997.

even extraordinary times to explain human viciousness. Numerous studies of cruelty show how men and women "like us" are capable of grotesque acts of violence against fellow human beings.'[18] This may sound cynical but perhaps the reasons why ordinary men and women behave with unspeakable savagery in wartime are not inexplicable and perplexing after all – this is how things are.

The propensity for cruelty is in all of us, and as illustrated by Christopher Browning's study of the German Reserve Police Battalion 101 on the Eastern Front, it rises to the surface for many when they are given complete authority over other human beings. Browning's pioneering study, *Ordinary Men*, demonstrated the readiness with which good neighbours could become brutally efficient killers and killing could become routine even for 'ordinary' men.[19] Browning reconstructs from the records of a West German court of inquiry the activities of the 500 'ordinary men' of the Reserve Police Battalion 101 who, within a period of 17 months from July 1942 to November 1943, shot and killed 38,000 Jews and jammed another 45,000 into freight cars to be shipped to the extermination camps. These were not the Eichman type of bureaucratic killers, issuing orders from behind their desks far from the killing fields, nor, with a very few exceptions, were they fanatical Hitlerites. They were working men in their thirties and forties, many of them police officers in civilian life, sent to Poland to do a job whose details were not disclosed in advance. When they were eventually told that their task was to murder Jews, nearly the entire battalion proceeded methodically and without apparent feeling. They killed their victims one by one, usually with a bullet to the base of the neck, including old men and women as well as children. To save time infants were not taken to the killing fields but murdered where they were found. When members of the battalion were interviewed by a German court in the 1960s nearly all of them claimed to have been 'horrified and disgusted' at first by what they were doing yet nearly all overcame their repugnance and joined the killing, even though their commanding officer offered to exempt without penalty those who chose not to take part. A handful accepted his offer but soon rejoined their comrades. The rest pitched in at once. Some murdered eagerly. A few found the work sexually exciting. Some even made jokes about it. For most the assignment quickly became uneventful. The men of Reserve 101 might have been painting a house or moving furniture around for all the emotion they displayed.[20]

What motivated the 'ordinary' killers of Reserve Battalion 101 were the same things that motivated the Hutus in Rwanda, or the Serbians in Srebrenica: group conformity, peer pressure, racism, societal norms and a sense of superiority. The implication of all this is that normal-looking individuals (who have spent most of their lives as civilians in peacetime) are innately violent, merely requiring the opportunities

18 Bourke, *An Intimate History*, p. 5.

19 See Christopher Browning, *Ordinary Men: Reserve Battalion 101 and the Final Solution in Poland* (New York, 1992).

20 This paragraph draws heavily on Jason Epstein's superb essay, 'Always Time to Kill', *The New York Review of Books*, 4 November 2004.

of war to revert to primordial savagery. Ervin Staub argues emphatically in his *The Roots of Evil: The Origins of Genocide and Other Group Violence*, that 'ordinary psychological processes and normal, common human motivations and certain basic but not inevitable tendencies in human thought and feeling' are the 'primary sources' of the human capacity for mass destruction of human life. It would be a lot easier and much more convenient to accept that the perpetrators are monsters and that their violence is not committed by ordinary people, rather than to agree with the view that 'evil arises out of ordinary thinking and is committed by ordinary people [and it] is the norm, not the exception.'[21]

Michael Mann, whose work has brought a new and impressive sociological rigour to the study of genocide, goes a step further arguing that although it is disturbing for us to acknowledge the ease with which ordinary people perpetrate the most astonishing atrocities, evil does not arrive from outside civilization nor from a separate realm we are tempted to call primitive. The sadistic torture and genocidal rapes, for instance, of thousands of Bosnian and Croatian women by Serb militias in the early stages of the Yugoslav war was not a peculiar feature of that conflict but a reality of warfare and the perpetrators behind such atrocities were not faceless primitive species of evildoers but real people with personalities and names.

Trying to keep good and evil separate from each other and away from real life is pointless, argues Mann, for evil is generated by civilization itself:

> Murderous ethnic cleansing has been a central problem of our civilization, our modernity, our concepts of progress and our attempts to introduce democracy. It is our dark side. Perpetrators of ethnic cleansing do not descend among us as a separate species of evildoers. They are created by conflicts central to modernity that involve unexpected escalations and frustrations during which individuals are forced into a series of more particular moral choices. Some eventually choose paths that they know will produce terrible results ... [As for] the rest of us [we] can breathe a sigh of relief that we ourselves have not been forced into such choices, for many of us would also fail them. Murderous ethnic cleansing comes from our civilization and from people, most of whom have not been unlike ourselves.[22]

What Mann calls the 'dark side' of human life, Michael Jackson calls, in his *In Sierra Leone*, a book about the brutality of the country's ten-year civil war, 'the business of reversing humiliation through violence.' Drawing on his experience before the war and supplementing it with post-war interviews Jackson helps us to understand those men and women who took to violence. In particular, he shows in convincing detail that atrocities were in fact the perpetrators' way of dealing with their own fear. Conflict can result from one being attacked but also from fear of being attacked. Unless one has been caught up in war, writes Jackson, and has:

---

21 Ervin Staub, *The Roots of Evil: The Origins of Genocide and Other Group Violence*, (Cambridge, 1989), p. 126.

22 Michael Mann, *The Dark Side of Democracy: Explaining Ethnic Cleansing*, (Cambridge, 2005), p. ix.

first-hand experience of the terror that comes of knowing that thousands of heavily armed individuals are bent on one's annihilation, it is hard to realize that most violence is not primarily motivated by evil, greed, lust, ideology, or aggression. Strange as it may seem, most violence is defensive. It is motivated by the fear that if one does not kill one will be killed, either by the enemy or by one's superiors. Against this constant anxiety, and the acute sense of fear and vulnerability that accompanies it, one conjures an illusion of power – torching buildings, shooting unarmed civilians, firing rocket grenades, smoking cannabis, shouting orders, chanting slogans, seeing oneself as Rambo, taunting, torturing, abusing the individuals one has taken captive. But all this display of might ... simply reveals the depth of one's impotence and fear.[23]

Human beings, like any other animal, will fight to the death when threatened or cornered. And once in battle, 'one defence is to distance the people on the other side. The person you maim or kill is not seen as someone as frightened as you, whose mother and father want him to come back from the war.'[24] This view of human relations is born out of the fact that psychological distancing and dehumanization of the enemy are quickly accomplished once the killing has started.[25] In the Mai Lai massacre, Private Varnado Simpson killed about 25 people: 'Men, women. From shooting them, to cutting their throats, scalping them, to ... cutting off their hands and cutting out their tongue. I did it.' When asked how he could bring himself in perpetrating such atrocities, Simpson said:

I just went. After I killed the child, my mind just went. And once you start, it's very easy to keep on. Once you start. The hardest – the part that's hard is to kill, but once you kill, that becomes easier, to kill the next person and the next one. Because I had no feelings or no emotions or no nothing. No direction. I just killed. It can happen to anyone.[26]

Distancing between perpetrator and victim is one of the keys to such behaviour and may be one reason why warfare is so savage and why war crimes and atrocities are now integral to the very prosecution of war. Speaking about his first murder of a Tutsi, a young Hutu by the name of Pio says:

I had killed chickens but never an animal of the stoutness of a man like a goat, or a cow. The first person, I finished him off in a rush not thinking anything of it, even though he was a neighbour, quite close on my hill. In truth it came to me only afterward: I had taken the life of a neighbour. I mean, at the fatal instant I did not see in him what he had been before; I struck someone who was no longer either close or strange to me, who wasn't exactly ordinary anymore ... like the people you meet every day. His features were indeed similar to those of the person I knew, but nothing firmly reminded me that I had lived

---

23 Michael Jackson, *In Sierra Leone* (Durham, North Carolina, 2004), p. 38.
24 Jonathan Glover, *Humanity: A Moral History of the Twentieth Century* (London, 1999), p. 50.
25 On psychological distancing and dehumanization see John W. Dower's brilliant *War Without Mercy: Race and Power in the Pacific War* (London, 1986), pp. 3-15.
26 Michael Bilton and Kevin Sim, *Four Hours in My Lai* (London, 1993), p. 7.

beside him for a long time … He was the first victim I killed; my vision and my thinking had grown clouded.[27]

We may never know precisely what motivated young Hutus like Pio to such bestial behaviour but it was the belief that 'the enemy was not a human being' (a belief cultivated by years of propaganda and social indoctrination) that turned minor differences into major and legitimized inhumane behaviour. In Nanking, the Japanese imperial army saw the Chinese as subhuman whose murder, rape would carry no greater moral weight than squashing a bug. In Rwanda, the enemy were 'cockroaches', in Vietnam, the enemy soldiers were 'gooks' and in former Yugoslavia 'vermin', 'insects' and 'dogs'.

But how does dehumanization overcome the evidence of common humanity? If one wants to get closer to the truth of what happened and why, it is as important to talk to the victimizers as to the victims. It not only the victims whose worlds one needs to enter, if one wishes to understand modern warfare, but also:

> the world of the gunmen, torturers, and apologists of terror. To such people, the idea that human beings are sacred rights-bearing creatures would be true only for their own. As concerns their enemies and their victims, they have carpeted together persuasive reasons for refusing to think of them as human beings at all. The horror of the world lies not just with corpses, not just with the consequences, but with the intentions, with the minds of the killers. Faced with the deep persuasiveness of ideologies of killing, the temptation to take refuge in moral disgust is strong indeed. Yet disgust is poor substitute for thought.[28]

So is it pointless to imagine a world where people are not hounded from their homes, starved to the death, tortured or massacred? A world where dignity prevails among warriors who choose to fight each other? What human history has taught us and continues to teach us is that for as long as there is war and human conflict, there will always be people – psychopaths and conformists, fanatics and opportunists, adventurers and moral cowards - willing to commit atrocities in exchange for a little power and privilege. And once you choose to look at the violence in this way, and notice that so much of it occurs in the familiar environment of everyday life, it no longer seems mindless, chaotic or medieval. There is a level on which violence has no reason or purpose – it exists to gratify itself. Nothing illustrates that point better than the abuses at Abu Ghraib as it will be shown in the next section.

Morality and Torture

Excesses and errors in foreign wars are common place; invading and occupying armies have always behaved badly. The Belgians in the Congo, the British in Kenya, the

---

27 Cited in a *Financial Times* review essay on Jean Hatzfeld's book on Rwanda, *Machete Season: The Killers in Rwanda Speak*, *Financial Times Magazine*, 16 July 2005.

28 Michael Ignatieff, *The Warrior's Honor: Ethnic War and the Modern Conscience* (London, 1988), p. 24.

French in Algeria, the Germans during the war, the Russians, the Japanese, the British, even the Dutch during the colonial wars, committed identical atrocities and practiced torture and sexual humiliation on despised, recalcitrant natives. But most did it before the age of the digital pocket camera. Images like the photograph of a naked Iraqi prisoner, cowering in front of barking dogs in Abu Ghraib, have forced us to come face to face with warfare's undiminished brutality and indiscriminate excess.

The late Susan Sontag, in a coruscating essay a few months after the Abu Ghraib scandal broke, expressed the notion that 'the photos are us.'

> You ask yourself how someone can grin at the sufferings and humiliation of another human being – drag a naked Iraqi man along the floor with a leash? Set guard dogs at the genitals and legs of cowering, naked prisoners? Rape and sodomise prisoners? Beat prisoners to death? – and feel naïve in asking these questions, since the answer is, self-evidently: people do these things to other people. Not just in Nazi concentration camps (and in Abu Ghraib when it was run by Saddam Hussein). Americans, too, do them when they have permission. When they are told or made to feel that those over whom they have absolute power deserve to be mistreated, humiliated and tormented. They do them when they are led to believe that the people they are torturing belong to an inferior, despicable race or religion.[29]

The real meaning of the pictures, insisted Sontag, is not that these acts were performed but that their perpetrators had no sense that there was anything wrong in what the pictures show. They were meant to be circulated for amusement. The problem, of course, runs deeper than the sad truth that some people take pleasure in the pain of others. The fact that the pictures were taken at all, never mind the cheerful expressions on the faces of the soldiers, suggests an atmosphere in which these soldiers had no reason to fear being punished for their behaviour.[30] The smiling faces of the tormentors confirm that as well as lacking moral judgment, these soldiers felt licensed to abuse.[31] And this is the crux of the matter.

Private Lynndie England – the female soldier in the case – did not think that there was anything wrong with holding a prisoner on a leash because her superiors had sanctioned it. Her character becomes irrelevant as soon as we realize that what made her deeds possible was a very particular political culture created in Washington DC in which almost anything went so long as it was deemed to serve the 'war on terror', as waged by the US.[32] This is where the abuse at Abu Ghraib has its origins. 'If White

---

29 Susan Sontag, 'What have we done?' in *The Guardian*, 24 May 2005.

30 On the new face of American war see Evan Wright, *Generation Kill: The Story of Bravo Company in Iraq – Marines who deal in bullets, bombs and ultra violence* (London, 2004); see also Anthony Swofford, *Jarhead: A Soldier's Story of Modern War* (London,, 2003).

31 Bowden, 'Lessons of Abu Ghraib'.

32 On the-end-justifies-the-means approach of the Bush administration in conducting their 'war on terror' strategy see Mark Danner, *Torture and Truth: America, Abu Ghraib and the War on Terror* (London, 2005); Karen J. Greenberg and Joshua L. Dragel (eds), *The Torture Papers: The Road to Abu Ghraib* (Cambridge, 2005); John Yoo, *The Powers of War and Peace: The Constitution and Foreign Affairs after 9/11*(Chicago, 2005).

House and Pentagon lawyers seek ways to circumvent the Geneva conventions,' wrote Ian Buruma in a trenchant essay for the *Financial Times* magazine, 'if torture is deemed permissible if the [US] president says so, if that same president divides the world into good guys and evil guys, and if it is unclear in Iraq who the enemies are, then it becomes hard to blame Lynndie England [and her fellow abusers] ... for playing pornographic games with real victims. The problem is not cultural, or personal but political ...'[33]

There was 'a before 9/11 and an after 9/11' as Coffer Black, the onetime director of the CIA's counter terrorist unit, put it in testimony to Congress in early 2002. Indeed. Soon after the 9/11 attacks, Americans 'took the gloves off' and began torturing prisoners and they have never really stopped. These words must by now be accepted as fact; not accusation. The problem stems in part from a moral equivalent that emerged after the events of 9/11 when America found itself under attack by a brutal, amorphous enemy which would go to extraordinary lengths and use whatever means it could to destroy the West. President George Bush repeatedly declared that the US were 'in a different kind of war' and had an obligation to defend itself. Yet, however 'different' that war was going to be, universal (and American) values required that civilized standards be maintained. A country as powerful as the US has many choices, even when struck by a blow as heavy as that of 9/11. Playing fast and loose with international law and the norms of civilized behaviour as the US administration chose to do after 9/11 was always going to be a self-defeating strategy.[34] In its casual disregard for international public opinion and cavalier approach to human rights, the United States has not only damaged the nation's moral standing but more crucially, undermined the very values that the 'war against terror' was supposed to encourage.[35] Francis Fukuyama may have exaggerated in the past with his 'end of history' assertions but he hardly did when he wrote about the 'seismic shift in the way much of the world [now] perceives the US, [and] whose image is no longer the Statue of Liberty but the hooded prisoner at Abu Ghraib.'[36]

---

33 Ian Buruma, 'Just following orders' in the *Financial Times Magazine*, 3 July 2004; See also Joanna Bourke, 'Torture as pornography' in *The Guardian*, 7 May 2004 and Katherine Viner, 'The sexual sadism of our culture, in peace and in war' also in *The Guardian*, 22 May 2004.

34 On America's flagrant violation of the established international codes of behaviour concerning prisoners of war see Philippe Sands' devastating critique *Lawless World: America and the Making and Breaking of Global Rules* (London, 2005); Michael Byers, 'A New Type of War: Blair's and Bush's attempt to change international law' in *London Review of Books* (6 May 2004); David Scheffer 'Terror suspects are entitled to legal protection' in *Financial Times*, 5 December 2005.

35 Graydon Carter, *What We've Lost: How the Bush Administration Has Curtailed Freedoms, Ravaged the Environment and Damaged America and the World* (London, 2004).

36 Francis Fukuyama, 'The Bush doctrine, before and after' in *Financial Times* 10 October 2005; see also Francis Fukuyama, 'Everything to regret', in *The Guardian*, 1 September 2005.

Nothing destroyed the moral case for the US war in Vietnam quite so effectively as the complicity of American forces in the use of torture. Of the many lessons of that conflict which optimists hoped the US had learnt, this was surely one of the most important. Why the US army in Iraq chose to ignore these lessons from Vietnam remains incomprehensible.[37] Years from now, the mistreatment of Afghan war detainees in Afghanistan, Guantanamo Bay and Abu Ghraib is likely to rank with the internment of Japanese civilians in World War II as a blot on the history of the United States.[38]

Modern wars have come to be defined by photographs. In Vietnam the photograph of the nine-year-old Phn Thi Kim Phuc running down a highway in agony and fear, her body on fire after a napalm attack, had a devastating global impact that still resonates several decades later. Nick Ut's photograph confirms the saying that one picture is worth 10,000 words. The searing image showed with terrible intensity how war maims and destroys innocent children and helped at the same time to galvanize anti-war sentiment in America, which in turn influenced US policy and led eventually to withdrawal.

The abuses at Abu Ghraib are a public rebuttal to Western claims that a rotten regime would be replaced with an enlightened democratic government in which human rights would be properly respected.[39] The purpose of this introduction is certainly not to provide an analysis of whether or not the occupation of Iraq turned out – to use the words of the liberal American magazine *The Nation* – 'to be a morally corrosive imperial adventure'[40] but it needs to be said that the sight of Americans torturing Iraqis in the same prison that Saddam Hussein used to torture Iraqis destroyed the credibility of the assumption that American and British troops would help to build a stable, representative democracy that respected human rights and the rule of law, 'a beacon' of liberty to the authoritarian regimes of the Middle East.[41] The reality, a few

---

37 See Anatol Lieven, 'A second chance to learn the lesson of Vietnam', in *Financial Times*, 8 June 2004; see also William B. Bader, 'From Vietnam to Iraq: Pretext and precedent' in *International Herald Tribune*, 27 August 2004.

38 See David Rose's *Guantanamo: America's War on Human Rights* (London, 2004), chapter 4 in particular; see also Jan Banning, 'Traces of war: Dutch and Indonesian survivors', www.openDemocracy.net, 19 August 2005. The photojournalist Jan Banning listens and portrays Dutch and Indonesian prisoners-of-war forced into labour and denied even minimal rights by their Japanese captors during the brutal Pacific war of 1941-45.

39 See Human Rights Watch report on torture in Iraq entitled 'Leadership Failure: Firsthand Accounts of Torture of Iraqi Detainees by US Army's 82nd Airborne Division', 25 September 2005. Available at hrw.org/reports/2005/us0905.

40 'Orders to Torture', *The Nation*, 7 June 2004; On America's Iraqi adventure see George Parker, *The Assassin's Gate: America in Iraq* (New York, 2005) and Andrew J. Bacevich, *The New American Militarism: How Americans Are Seduced by War* (New York, 2005); see also a polemical essay by Tony Judt entitled 'The New World Order' in the *New York Review of Books*, 14 July 2005.

41 For an equally polemical take on the Iraq war see John Gray's essay 'Power and Vainglory: Iraq isn't another Vietnam – it's much worse' in *The Independent*, 19 May 2004;

months after the third anniversary of the day US and British forces began to advance on Baghdad, could not be more different.[42]

Now that we know all about the hidden prisons and the torture chambers and the beatings, is there a moral difference between Saddam Hussein's behaviour and the behaviour of American and British troops? As the months of the Iraq war have turned into years, Iraqis, to use the bitter words of Iraqi cleric Sheik Mohammed Bashir, have discovered that:

> freedom in this land is not ours. It is the freedom of the occupying soldiers in doing what they like … abusing women, children, men, and the old men and women who they arrested randomly and without any guilt. No one can ask them what they are doing, because they are protected by their freedom … No one can punish them, whether in our country or their country. They expressed the freedom of rape, the freedom of nudity and the freedom of humiliation.[43]

When morality vanishes from the battlefield, a war can never be won. It does not matter a great deal whether you choose to call this kind of treatment abuse or some other euphemism, the point remains that torture, like any other atrocity lives on in the mind of the tortured. And as any abused prisoner at Abu Ghraib, Guantanamo or Afghanistan[44] will tell you, every time a person is reduced to a howling beast by deliberately inflicted pain, our civilization crumbles a little more, until in the end there will only be barbarism.

Only in the pages of bad spy novels should governments throw people they dislike in stateless prisons and subject them to torture, abuse and humiliation.[45] *Enemy Combatant: A British Muslim's Journey to Guantanamo and Back*[46] is not the title of a bad spy novel but, in fact, the story of what happened to Moazzam Begg, a British-born Muslim, when the US administration decided that he was an 'enemy combatant' in the 'war on terror'. Begg, the son of a Pakistani banker who

---

see also David Runciman, *The Politics of Good Intentions: History, Fear and Hypocrisy in the New World Order* (Oxford, 2006).

42 As Rupert Cornwell put it in a newspaper opinion piece: 'It has taken more than three years, tens of thousands of Iraqi and American lives, and $200bn of treasure – all to achieve a chaos verging on open civil war' in the *Independent*, 9 March 2006; For a gritty and compelling firsthand account of post-conflict Iraq see Mark Etherington's *Revolt on the Tigris: The Al-Sadr Uprising and the Governing of Iraq* (London, 2005).

43 Sheik Mohammed Bashir, Friday prayers, Um al-Oura, Baghdad, 11 June 2004. Cited in Edward Coy's *Washington Post* article 'Iraqis put contempt for troops on display', 12 June 2004.

44 On the America's secret network of Afghan jails see Suzanne Goldenberg's special report for the *Guardian*, 23 June 2004.

45 See Eugene Robinson's extremely perceptive opinion piece for the *Washington Post*, 4 November 2005.

46 See Moazzam Begg, *Enemy Combatant: A British Muslim's Journey to Guantanamo and Back* (London, 2006). See also Joseph Lelyveld's essay, 'Interrogating Ourselves' in *New York Times*, 12 June 2005.

migrated to Birmingham and a former law student at the university where I teach, was arrested in Pakistan in 2002, where he was helping set up education programs for children, in the panic-stricken months after the 9/11 attacks. He was then sent to Guantanamo where he spent three years in prison, much of it in solitary confinement, and was subjected to over 300 interrogations, death threats and torture, witnessing the killings of two detainees. He was released early in 2005 without explanation or apology. As Stephen Fidler put it in a review essay of Begg's book: 'One does not have to believe Begg is as innocent as he says to deplore what befell him. He was shackled, beaten, threatened with death and isolated for three years on the basis of evidence that would not have been given the time of day by a court of law.'[47]

At the start of the twenty-first century we should not be debating the use of torture. Questions such as whether torture works and whether the Geneva conventions still hold should not have become central moral issues of our own age. No argument is more clichéd than that which asks why we should adhere to the Geneva conventions when our terrorist enemies do not.[48] Torture – no matter how 'light' and whatever the pro-torture brigade argues – cannot be justified on any grounds.[49] For it is a sign of desperation, an admission that your side has no other resources left. What is more, torture does not actually work: prisoners treated like those in Abu Ghraib will confess to anything. Torture is unjustified and utterly unacceptable because it is wrong to inflict pain on defenceless captives, because it breaks the international conventions and because the costs outweigh the benefits of any intelligence it may elicit. In the final analysis, torture is a weapon of punishment and a terrorist method: not a path to democracy or freedom.[50]

And if all of the above arguments are not enough there is another, even stronger one for drawing a firm line against any use and form of torture. It poisons and brutalizes the society which allows it to happen. The days of France's bloody colonial war in Algeria may be long past, but its reverberations continue to have a profound impact on French society. Alistair Horne, wrote in his *A Savage War for Peace* that the use of torture in Algeria became for France a growing cancer, leaving behind a poison which would linger in the French body politic long after the war had ended.[51] Roughly a million and a half Frenchmen, including the President of the Republic, Jacques Chirac, fought in Algeria, and about a quarter of them are believed by the French health authorities to suffer from psychological traumas. The French military, unsurprisingly, still refuses to accept that the widespread use of torture was anything

---

47 See *Financial Times magazine*, 18 March 2006.

48 For a diametrically opposite view see, for example, Max Boot, 'Necessary Roughness: Terrorists don't rate Geneva protections' in *Los Angeles Times*, 20 January 2005.

49 See Anthony Lewis, 'Making Torture Legal' in *New York Review of Books*, 15 July 2004. See also Stanley Cohen, 'Post Moral Torture: From Guantanamo to Abu Ghraib' in *Index on Censorship's* special issue on torture (Vol. 34, No.1, 2005), pp. 24-30; James Clasper, 'The Rotten Tree of Torture' in *The Liberal* (February-March 2005).

50 See David Rose, 'Using Terror to Fight Terror' in *The Observer*, 26 February 2006.

51 Alistair Horne, *A Savage War for Peace* (New York, reprinted, 2006); see also Adam Shatz essay 'Torture of Algiers' in *New York Review of Books*, 21 November 2002.

other than a necessary means of fighting a just war but if torture inspires widespread condemnation in France today, as it failed to do during the Algerian war, it is mainly because one can no longer defend it as an unfortunate necessity. Numerous recent studies on the Algerian war show that French violence in Algeria was designed to terrify, subdue and exhibit power rather than to extract information.[52]

Ultimately, the images of Abu Ghraib are a cause of shame for all of us because they show that barbaric and atrocious behaviour needs little encouragement to flourish. All that is needed is a sense that the ordinary rules do not apply. These snapshots tell us exactly what we need to know about who we are and the tissue-thinness of our veneer civilization. Because the maltreatment of Iraqi detainees and Muslim 'terrorists' dehumanizes not just those who man the cages and take the callous photos, but also the societies in whose name the cages were built and the guards recruited. The moral values of governments bear directly on the reputations of all citizens. 'It's not about who our enemies are,' said Senator John McCain, a man who has felt the outrage of torture on his lacerated skin and broken bones in a Vietnamese prison, 'It's about who we are.' In other words, it is depraved to take pleasure from the ill-treatment or torture of fellow human beings and societies which permit or encourage such behaviour are inferior moral societies.

The Abu Ghraib snapshots also tell us, that despite the profound and largely positive socio-cultural changes that have taken place in Western society since 1914, barbarity, atrocity and terror still loom large.[53] We have two kinds of motives to look into history. There is curiosity about the past, what happened, who did what, and why; and there is also the aim of understanding the present and how to place and interpret our own times, experiences and hopes for the future. What links twentieth century warfare with the current history of conflict in the 'war against terror' is the notion that warfare is a state of being, not a state of conflict. People do have a choice even if they have opted to march into battle. In the same way there are just wars and unjust wars, forms of killing that are necessary and forms that shame us all, there are human and inhuman warriors. To say this is not to establish a hierarchy of suffering, nor to downgrade the death and destruction of others but the premeditated killing of non-combatants whether in Kosovo, Vietnam or Chechnya is different from acts committed in the heat of battle.

Although no one could seriously claim that the history of twentieth century warfare has been neglected by scholars, only a small number of writers have tried to deal directly with barbarity in warfare, examining the rationale, motives, ideology, and moral resources of people in wartime. The principal aim of the essays which follow in this collection is to understand how and why this degeneration of human

---

52 See Raphaelle Branche, *La Torture et l'Armée Pendant la Guerre d'Algérie* (Paris, 2002); Sylvie Thénault, *Une Drôle De Justice: Les Magistrats dans la guerre d'Algérie* (Paris, 2002) and Irwin M. Wall, *France, the United States and the Algerian War* (California, 2002).

53 See the introductory chapter of Niall Ferguson's *The War of the World: History's Age of Hatred* (London, 2006).

behaviour happens and to attempt to offer explanations for its recurrence. Across the bleak landscape of the twenty-first century, in an age when actual war, violence and torture are becoming addictive forms of entertainment, it is now more critical than ever to deepen our understanding of the brutality and inexcusable sadism that can be displayed by human beings in times of war.

PART I
Stories of Atrocity

Chapter 2

Barbarism in War: Soldiers and Civilians in the British Isles, c.1641–1652

Frank Tallett

One important indicator of barbarity in war is the mistreatment of civilians by soldiers. This essay explores the use of violence in the Early Modern period by applying this indicator to the struggles which took place in the British Isles during the Wars of the Three Kingdoms.[1] These conflicts provide an especially appropriate historical laboratory to observe the use of violence, first because of their geographical spread and second because of their timing. Geographically, the fighting ranged across and between the kingdoms of England, Ireland and Scotland. Charles I led two unsuccessful campaigns against Scotland in 1639 and 1640. The following year saw the outbreak of a rebellion by Irish Catholics against the English and Scots settlers, and shortly thereafter there were four sets of forces fighting in Ireland: a Scots army, troops loyal to the king, those loyal to Parliament and the Catholic Irish forces usually referred to as the Confederates. The definitive conquest of Ireland by Oliver Cromwell's English army came ten years after the outbreak of the rebellion. In England, the first civil war between king and Parliament lasted from 1642 to 1646. In September 1643 the king agreed to a cessation of hostilities with the Confederates in order that a force of former Irish rebels could be brought to England to assist him in his war with Parliament, and between October 1643 and June of the following year, almost 23,000 men were introduced from Ireland.[2] In 1644, Parliament also secured outside assistance from a Scots army under the leadership of the seasoned veteran, Alexander Leslie. Second and third civil wars occurred in 1648 and 1650 respectively. On both occasions, Scots forces unsuccessfully intervened on the side

---

1   Historians have recognized that the term *English* civil wars is inadequate but have struggled to find a name for these conflicts, though most accept the need for some non-anglocentric epithet. The 'Wars of the Three Kingdoms' has found some favour. See R. Armstrong, *The 'British' of Ireland and the Wars of the Three Kingdoms* (Manchester, 2005) for the term's most recent use and M.J. Bennett, *The Civil Wars in Britain and Ireland, 1638-1651* (Oxford, 1997), pp. x-xii for a discussion of the issues behind the nomenclature.

2   In fact, the first batch of soldiers to arrive in October 1643 comprised Protestants who had been fighting the Catholic rebels; and overall about half of the soldiers were native Irish: J.L. Malcolm, 'All the king's men: the impact of the Crown's Irish soldiers on the English Civil War', *Irish Historical Studies*, 83 (1979), p. 252.

of the royalists, and Scottish independence was effectively ended by the campaigns of the New Model Army. Meanwhile, Scotland had undergone its own internal wars, reflecting the animosities of Lowlanders and Highlanders and differing clan loyalties, in which Irish troops played a considerable part.[3]

All of these conflicts were closely interrelated, yet the fighting in each theatre of war retained distinctive characteristics. This was especially so with regard to the nature of the fighting, the regulation of soldiers' behaviour, the levels of discipline and the attitude of the troops towards the civilian populations. And, as we shall see, the nature, level and incidence of violence used by soldiers against civilians also varied quite dramatically. By adopting a comparative approach to these conflicts it is possible to shed light on the factors governing the treatment of civilians by soldiers, and to draw some conclusions which are applicable not just to the British Isles during the 1640s, but to Europe during the early modern period and beyond.

It might be objected that the British conflicts were not typical of warfare of the period. In a bellicose epoch when warfare was becoming increasingly sophisticated, England in particular has a reputation as a pacifistic state, and the British Isles are frequently regarded as isolated from military developments on the continent, and hence untutored or backward in military science. Yet such characterizations do not accord with reality. To be sure, only Ireland had recent and extensive local experience of conflict, under Elizabeth: until the Bishops' Wars of 1639-40, there had been no serious fighting on Scots or English soil since the middle of the sixteenth century.[4] Sir Henry Slingsby noted in 1639 that military exercises 'are strange spectacles to this nation, in this age that have lived thus long peaceably, without noise of drum or shott, and after we have stood neuters, and in peace, when all the world hath been in armes,'[5] while Walter Meredith in 1642 likened the English nation to 'blades still kept sheathed, rusted in their scabbards; for to say truth, they were grown weake and effeminate for lack of imployment.'[6] Yet such remarks were a reflection of fears about the dangers of civil war rather than an objective analysis of the country's military capacity. In reality, there existed a high degree of military expertise in all three kingdoms of the British Isles on the eve of the wars. Fundamental to this was the experience gained by the stream of recruits who fought in the continental wars of the period. On average, 7,000 Englishmen had been raised annually by the Crown for overseas service between 1598 and 1602, with approximately 3,000 others joining of their own accord, and the pace

---

3   Two Irish regiments joined the royalist forces of the Marquis of Montrose in June 1644 and were crucial to his success: J. Lowe, 'The Earl of Antrim and Irish Aid to Montrose in 1644', *Irish Sword*, 4 (1960), pp. 191-8; C. Danachair, 'Montrose's Irish Regiments', *Irish Sword*, 4 (1959), pp. 61-7.

4   An English army had defeated the Scots at Pinkie in 1547; 9,000 men had been mobilized to crush rebellion in England in 1549; and sizeable Scots and English forces had been involved in the uprising against the regency of Mary of Guise in Scotland in 1559-60.

5   *Original Memoirs Written During the Great Civil War* (Edinburgh, 1806), pp. 22-23.

6   Thomason Tracts (hereafter cited as TT), E109 (8) *The fidelity, Obedience and Valour of the English Nation*, August 1642.

of recruitment did not slacken thereafter.[7] Four English regiments were in continuous service with the United Provinces from 1600 until at least 1632, seven in 1626.[8] And some 10,000-15,000 Englishmen volunteered for service in the Thirty Years War.[9] The Scots and the Irish similarly provided considerable numbers of recruits. Three Scottish regiments were in Dutch service in 1628, and as many as 20,000 Scots may have fought in Germany in the 1630s.[10] Slingsby observed that the Scots 'are become most warlike, being long experienced in the Swedish and German wars.'[11] An Irish regiment had been formed to serve under Archduke Albrecht in the Low Countries and overall up to 10,000 people moved from Ireland to the Low Countries between 1586 and 1622, most of them enlisting in the Spanish Army of Flanders.[12] Large numbers of the population of the British Isles had thus been tutored in the continental 'schooles of warre' in the decades preceding the outbreak of the civil wars, and many of these soldiers returned to fight in their home-grown conflicts after 1642, providing a solid cadre of veterans in the armies.[13] Moreover, the practical experience of military matters gained by serving soldiers was supplemented by a burgeoning literature on the subject of war. This ranged from highly specialized drill books, through maps, memoirs and military narratives, to propaganda pieces. The first half of the seventeenth century especially witnessed an outpouring of 'atrocity' material, portrayed in the engravings of Jacques Callot, Lukas Kilian and Ulrich Franck as well as in books and newsletters, such as the *Swedish Intelligencer*, much of which focused upon the horrors of the Thirty Years War.[14] This literature not only helped educate intending and practicing soldiers, it also raised public awareness of military matters.[15] One should therefore

---

7   P.E.J. Hammer, *Elizabeth's Wars. War, Government and Society in Tudor England, 1544-1604* (London, 2003), pp. 245-48.

8   Albert Joachim, ambassador, to States General, 25 August 1632, printed in 'Papers illustrating the history of the Scots brigade in the service of the United Netherlands, 1572-1782', *Scottish History Society*, 32 (1899), p. 410; and D. Trim, 'Fighting "Jacob's Warres". The Employment of English and Welsh Mercenaries in the European Wars of Religion: France and the Netherlands 1562-1610', unpublished PhD thesis (University of London, 2003), p. 340.

9   C. Carlton, 'The Face of Battle in the English Civil Wars', in M.C. Fissel (ed.), *War and Government in Britain, 1598-1650* (Manchester, 1991), p. 229.

10   *Scottish History Society* 32 (1899), xii; J. Kenyon, *The Civil Wars of England* (London, 1988), p. 44.

11   *Memoirs*, p. 23.

12   J. Casway, 'Henry O'Neill and the Formation of the Irish Regiment in the Netherlands, 1605', *Irish Historical Studies*, 28 (1973), pp. 481-88; G. Henry, *The Irish Military Community in Spanish Flanders, 1585-1621* (Dublin, 1992), p. 145.

13   Carlton, 'Face of Battle', p. 229; R. Loeber and G. Parker, 'The Military Revolution in Seventeenth-century Ireland', in J.H. Ohlmeyer (ed.), *Ireland from Independence to Occupation, 1641-1660* (Cambridge, 1995), pp. 72-3 on the Irish returnees.

14   As Barbara Donagan, 'Halcyon Days and the Literature of War: England's Military Education before 1642', *Past and Present*, 147 (1995), p. 75, notes: 'It is indeed hard to exaggerate English fascination with and horror at the German example.'

15   For a listing of the books see M.J.D. Cockle, *A Bibliography of Military Books up to 1642* (London, 1957) and H.J. Webb, *Elizabethan Military Science: The Books and the*

guard against any kind of British exceptionalism when discussing the conflicts of the 1640s, which draws a distinction between amateur scuffles in Britain and professional wars on the continent.[16]

A study of the conflicts in the British Isles is also appropriate because of their timing. As Geoffrey Parker notes, by the mid seventeenth century, most of the rules and conventions which restrained the conduct of soldiers were in place.[17] What did these comprise? To start with, there were the articles of war published by the army commander at the start of a campaign. These were rarely drawn up afresh, generals instead preferring to amend existing examples. Thus, Maurice of Nassau's *Artikelbrief* of 1590 served as a model for Sweden's Gustavus Adolphus and informed the codes of the English New Model army and Leslie's Scottish forces.[18] The codes regulated the soldiers' conduct with civilians, typically by forbidding looting, theft and rape, which all carried the death penalty.[19] Soldiers could not plead lack of awareness of the articles of war, for printed copies were given to new recruits and read out at general and weekly assemblies of the army 'so that no man pretend ignorance.'[20] Moreover, they had become easier to memorize by the 1640s, as offences were recorded under a series of subject headings rather than being listed piecemeal.[21]

Offenders could be tried by court martial for offences against the articles of war. There was no formal legal and punitive mechanism to back up the so-called

---

Practice (London, 1965) for an earlier period; Donagan, 'Halcyon Days and the Literature of War', pp. 72-100.

16 As P.R. Newman, *The Old Service. Royalist Regimental Colonels and the Civil War, 1642-46* (Manchester, 1993), p.8, notes 'Historians who insist upon the specious argument that the Civil War was fought by gentlemen amateurs, and was therefore in some degree less intensively prosecuted than the wars of Europe, have missed the point by a mile.'

17 G. Parker, 'Early Modern Europe' in M. Howard et al (eds), *The Laws of War: Constraints on Warfare in the Western World* (New Haven, 1994), p. 41.

18 The *Artikelbrief* is re-printed in J.W. Wijn, *Het krijgswezen in den tijd van Prins Maurits* (Utrecht, 1934). An English translation by Henry Hexham appeared in 1643. Adolphus' articles were translated by, among others, Robert Ward, *Anima'dversions of ware; or a militarie magazine of the truest rules, and ablest instructions, for the managing of warre* (London, 1639), bk ii, pp. 42-54. On the dissemination of the codes, see C.H. Firth, *Cromwell's Army. A History of the Soldier during the Civil Wars, the Commonwealth and the Protectorate* (London, 1962), pp. 279-81; G. Oestreich, *Neostoicism and the Early Modern State* (Cambridge, 1982), p. 125; M. Roberts, *Gustavus Adolphus. A History of Sweden, 1611-1632* (2 vols, London, 1953, 1958), vol II, p. 240.

19 See, for example, TT E116 (34), 8 September 1642, *Laws and Ordinance of Warre*, issued by the Earl of Essex: 'Rapes, ravishments, unnaturall abuses shall be punished with death; theft and robberie exceeding 12 pence shall be punished with death; noone ... shall waste, spoile or extract any victuals, monie ... upon pain of death; to take a horse out of plough or wrong a husbandman in their person or goods ... shall be punished with death.'

20 TT E109 (38) *Camp Discipline or the Souldier's Duty. Articles and Ordinances of Warre. To be Observed by the Armie of Scotland*, 10 August 1643.

21 B. Donagan, 'Codes and Conduct in the English Civil War', *Past and Present*, 118 (1988), pp. 83-4.

laws of war, a set of long-standing, internationally recognized conventions, which also governed a soldier's conduct. For their enforcement, these relied on a sense of professional honour and the threat of reciprocal action if they were breached.[22] Some knowledge of the conventions of war was general among both soldiers and civilians, but could not always be assumed. Thus the townsmen of Bradford misunderstood a captured royalist officer who 'cried out for quarter, and they poor men not knowing the meaning of it, said, – "aye, they would quarter him," and so killed him.'[23] If the conventions for dealing with military prisoners were not always well understood, the laws of war gave only an ambiguous protection to civilians. Although they distinguished between combatants and non-combatants, this distinction was not absolute, and the conventions allowed for numerous occasions on which non-combatant status might be overridden: it was legitimate to destroy civilians' property in pursuit of a scorched earth policy; when a town was besieged the attackers had the right to keep the inhabitants inside; and it was permissible to sack a town which had unreasonably refused to surrender. Such actions were not praiseworthy, but they were excusable under the laws of war. Perhaps most importantly, the laws of war distinguished between actions committed in the heat of the moment, which could be forgiven, and those committed in cold blood, which could not.[24] Thus, a royalist preacher warned his listeners that they should 'neither do, nor ... suffer to be done, in coole blood, to the most impious Rebells, any thing that savours of immodesty, Barbarousnesse, or inhumanity.'[25]

The Contexts of Violence in the Three Kingdoms: Supply, Disorder and Siege Warfare

If we turn, then, to look at the conduct of the wars in England, Ireland and Scotland in detail, it is possible to discern some common themes, despite the significant differences which also existed. Wherever the theatre of conflict, violence towards civilians was highly contextualized. First, much of the violence occurred

---

22 Though there was a melding of the laws and articles of war in the regulations governing the Scots Army of 1643. The conclusion indicated that anything not mentioned in the individual articles would be 'judged by the common customs and constitutions of warre', a handy catch-all phrase: TT E109 (38) *Camp discipline ... to be observed by the Armie of Scotland*, 10 August 1643.

23 T. Wright (ed.), *The Autobiography of Joseph Lister* (London, 1842), p. 17, cited in Donagan, 'Codes and Conduct', p. 82.

24 For the persistence of such attitudes into the twentieth century, see the passage in the *Australian Official History of the Great War* excusing the bayoneting of men attempting to surrender on the grounds that 'such incidents are inevitable in battle'. Quoted in Keegan, *The Face of Battle. A Study of Agincourt, Waterloo and the Somme* (Harmondsworth, 1983), p. 47 and subsequent discussion.

25 Edward Symmons, *A Military Sermon* (Oxford, 1644), p. 35 quoted in B. Donagan, 'Atrocity, War Crime and Treason in the English Civil War', *The American Historical Review*, 99 (1994), p. 1144, and pp. 1140-48 for an excellent discussion of the issues.

as part of the soldier's attempts to secure the necessities of life: food, shelter, fuel, clothes and money in his pocket. Because the central pay and supply systems of the armies were poorly organized and inefficient, soldiers were obliged to live off the resources of the locality and its population.[26] They achieved this in a number of ways, which all involved the use or the implicit threat of violence. Troops were billeted in local households; supplies were requisitioned at prices set by the army, or simply taken by force; and taxes, known euphemistically as 'contributions', were levied. True, commanders tried to ensure that this was done in a controlled manner, with as little damage as possible: the theft of civilian property was forbidden, householders were to be compensated for requisitioned goods and soldiers were meant to pay for their board and lodging at established rates.[27] Yet the expectation that soldiers would pay for what they took was unrealistic, since wages were not only low, but were paid infrequently or not at all.[28] In these circumstances, the soldiers simply took what they needed, using force when necessary. Thomas Stockdale complained of the 'abuse of the soldiery, under which this part of Yorkshire now groans,' but admitted that if only the troops were properly paid, 'the sufferance and wrong would be unto many less sensible.'[29] The system of living off the locality was in any event a standing invitation to the disorderly elements in an army to abuse civilians. Thus Henry Slingsby recorded that during the march to Oxford in 1644, 'We had in our company soldiers so unruly that they gave the whole country an alarum against us; they would ride out on every hand, rob the carriages, and play such pranks as we could expect no less than to be met with by the enemy.'[30] Even more seriously, the lack of pay and supplies might cause a general breakdown of discipline. As one contemporary observed, 'The greatest weakening of an army is disorder [indiscipline]. The greatest cause of disorder is want of pay.'[31] A collapse of discipline was invariably followed by attacks on civilians and uncontrolled pillaging of their houses, farms, crops and orchards.

It was not just when soldiers were billeted upon civilians, or when they scavenged for food and plunder, that violence occurred. The imposition of a siege was a second, though less frequent, danger point for civilians. The inhabitants of a town might

---

26 The exception to this was Cromwell's army of conquest in Ireland. See below.

27 See the schedule of allowances for the Scots Covenanting army of 1644 agreed by the Committees of Both Kingdoms to 'vitiate the evils' of billeting: TT E30 (16) *Declaration of the Committees for the Billeting of Soldiers* 1643. See also J.M. Rushworth, *Historical Collections of Private Passages of State* (6 vols, 1659), vol V, p. 612ff.

28 Kenyon, *The Civil Wars*, pp. 125-27.

29 Thomas Stockdale to Fairfax, 5 March 1640, in Fairfax, *The Fairfax Correspondence. Memoirs of the Reign of Charles I*, ed. G.W. Johnson, (2 vols, London, 1848), vol I, pp. 203.

30 *Memoirs*, p. 58.

31 TT E116 (36) *Observations Concerning Princes and States upon Peace and Warre*, September 1642, observation 16. In similar vein, Thomas Fairfax wrote to his father that, 'Our soldiers have been put to hard service and strict obedience; but if they want pay, both these will be neglected': *Fairfax Correspondence*, vol I, p. 245.

be forced by the garrison to help with the digging of earthworks.[32] More seriously, the besieging forces might refuse to allow civilians to leave a town so that they ate up the garrison's provisions. At Colchester, for example, women who slipped out were stripped and driven back into the town.[33] Captain Hodgson recalled that at the siege of Pontefract in 1645, 'They began to turn out women and children and an old man; and … Colonel Overton … sent them in again.'[34] At Limerick, those who tried to escape were 'whipped back into town', and when this proved an insufficient deterrent, 'a gibbet was erected in the sight of the walls, and 1 or 2 persons hanged up … and by this means they were so terrified that we were no further disturbed on that account.'[35] Civilians trapped in a siege ran the additional risk of being caught by artillery fire from the besieging forces. In practice, the danger seems to have been slight, as the example of Limerick suggests. The 3,000 cannon balls and 500 mortars fired by Ireton's army between June and August 1650 did little harm, 'causing not five pounds of hurt nor killing 7 men.'[36] The siege of Bradford by the royalists in 1644 'frighted many, but killed few',[37] while one week's battering of Wardour castle 'had done little other hurt save only to a chimney-piece, by a shot entering at a window.'[38] Most dangerous for civilians was the sack which followed the successful storming of a town. It was not just that the victorious troops believed themselves entitled to plunder the defeated town, but the psychological relief at having survived the dangers of the attack led them to commit excesses. Thus at the end of the siege of Leicester in May 1645, the royalist infantry murdered prisoners and 'put diverse women inhumanly to the sword'; civilians were similarly killed in the final assault on Basing House by Cromwell's troops in the same year; while at Carregmayne castle, south east of Dublin, they 'slew both man, woman and child' following the storming of the town.[39] It goes without saying that civilians who had helped with

---

32 P. Styles, 'The city of Worcester during the civil wars, 1640-60', in R.C. Richardson (ed.), *The English Civil Wars. Local Aspects* (Stroud, 1997), p. 200; J. Adair, *By the Sword Divided. Eyewitnesses of the English Civil War* (London, 1983), p. 77 on Oxford.

33 Kenyon, *The Civil Wars*, p. 185. The water supply had earlier been cut off and the lead pipes melted down for use as bullets. See also TT E451 (28) *A Letter from the Leaguer at Colchester*, 8 July 1648.

34 *Memoirs of Captain John Hodgson*, in Slingsby, *Memoirs*, p. 105.

35 E. Ludlow, *The Memoirs of Edmund Ludlow, Lieutenant-General of the Horse in the Army of the Commonwealth of England, 1625-1672,* ed. C.A. Firth (2 vols, Oxford, 1894), vol I, p. 284. J.T. Gilbert, *A Contemporary History of Affairs in Ireland from 1641 to 1652* (3 vols, Dublin, 1879-80), vol III, p. 440, suggests that escaping civilians were put to the sword.

36 Lord Muskerry to Ormonde, the royalist commander in Ireland, quoted in J. Burke, 'The New Model Army and the Problems of Siege Warfare, 1648-51', *Irish Historical Studies*, 27 (1990), p. 20.

37 *Memoirs of Captain John Hodgson*, p. 99.

38 Ludlow, *Memoirs*, vol II, p. 51. See also TT E42 (26) *Mercurius Aulicus*, 30 March 1644, which suggested that a mere 35 people were killed in the siege of Newark, although 1,000 large cannon balls had been fired.

39 C. Carlton, 'The Impact of the Fighting', in J. Morrill (ed), *The Impact of the English Civil War* (London, 1991), pp. 30-31; TT E141 (27) *A Letter Sent from Chester*, 2 April 1642.

the defence of a town could expect little mercy. Thus when the castles of Knock and Blackwood fell, the women and children who had thrown stones from the ramparts suffered: 'the souldiers being enraged against them … quartered both women and children excepting some five.'[40]

England and the Restraint of Violence: Rules, Regular Armies, and Brothers in Arms

If there are common themes to be seen, there were also significant differences in the treatment of civilians by soldiers across the three kingdoms. It is clear that the level and nature of violence was most restricted in England. This is not, of course, to suggest that the wars here were always conducted humanely. The deliberate murder and rape of civilians was certainly not unknown, particularly in the context of the storming of a town. In one of the worst massacres, up to 1,800 soldiers and townspeople died in the sack of Bolton by colonel Broughton's troops in May 1644, with no quarter given to soldiers or civilians. Eyewitnesses compared the event to the sack of Magdeburg in the Thirty Years War which had gone down as a byword for military brutality.[41] Yet, on the whole, the violence was restricted to verbal abuse, bullying, beatings and sexual harassment, while murder remained relatively unusual, torture even more so. How is this relatively restricted level of violence to be explained? One key reason was the recognition early on that this was not a rebellion but a regular war, and hence should be fought according to the normal rules of war which gave some protection to non-combatants. To be sure, royalists referred to their opponents as 'rebels',[42] their opponents referred to the royalists as 'traitors', and much of the discourse on both sides was extreme. Typical was the parliamentarian supporter who wrote of the 'malevolent and ill affected spirits around [the king] … that desire to swimme through a sea of blood.'[43] There was a crisis at the start of the conflict when the royalists threatened to execute three prisoners for treason. But when Parliament threatened to do the same by way of reprisal the issue was settled, and thereafter the normal rules of war were, by and large, observed. What Barbara Donagan has called the 'principle of beneficial mutual restraint' – the equivalent of the twentieth century's 'mutually assured destruction' – was at work here.[44] Moreover, restraint was encouraged by the fear that if limits were not set to the violence, then England would descend into the spiral of unrestricted mayhem that contemporaries had read about in the Holy Roman Empire. The 'German lesson' thus proved important in the restriction of violence. It was in this context that Prince Rupert argued that if the laws

---

40 TT E116 (24) *Exceeding Happy News from Ireland*, 9 September 1642.

41 F. Holcroft, *The English Civil War around Wigan and Leigh* (Wigan, 1993), esp. p. 26.

42 See for example, the sermon preached before Prince Maurice's army, at TT E53(19) *Militariae sermon wherein is discovered the nature and disposition of a rebel*, 19 May 1644.

43 TT E109 (21) *A New Discovery of the Design of the French*, 21 August 1642.

44 Donagan, 'Atrocity, War Crime and Treason', p. 1140 and passim for a penetrating discussion of the issues in respect of England.

of war were not observed, then 'the English nation is in danger of destroying one another or … of degenerating into such an animosity and cruelty that all [elements] of charity, compassion and brotherly affection, shall be extinguished.'[45]

Enforcement of the rules of war was made easier by the fact that troops were organized from the outset into regular armies, commanded and officered by men with previous military experience. At the top levels of command, Cromwell and Fairfax were notable disciplinarians; while in the army of the Earl of Essex, for example, Captain Douet was not alone in regularly haranguing his soldiers on the need to pay for anything they requisitioned and 'to use no uncivill language, especially to a woman or maid.'[46] To be sure, there were exceptions to these examples of military professionalism. Major Purefoy and Colonel Bridges, garrison commanders at Compton and Warwick respectively, gained unsavoury reputations for colluding with their men in the commission of disorders.[47] And even well-intentioned officers sometimes struggled to have themselves obeyed. Robert Duckenfield admitted that despite his men's pillaging, 'I dare not take in hand to punish them'; while Nehemiah Wharton's admittedly sanctimonious letters reveal the inability of Colonel Essex to enforce a prohibition on plunder.[48] Even Cromwell, faced with a lack of supplies for his army, found it hard to enforce his order during the 1650 Scottish campaign that civilians 'should be protected from the plunder and violence of the soldiers', and the area around Leith and Edinburgh was so ravaged that 'it was sad to see the devastation that was made.'[49] Moreover, as the wars went on, the difficulties of enforcing discipline began to mount, particularly from the autumn of 1643. Recruits in the first two years of the war were generally well motivated, but as their numbers were eroded by wounds and disease, there was an increased resort to impressment by both sides. What this produced was the dregs of society, men who were idle and dissolute in civilian life and whose habits did not change when they were enrolled. The Suffolk levies of 1643 frightened even Cromwell: 'They are so mutinous that I may justly fear they would cut my throat,' he wrote.[50] Yet even when allowance is made for the unwillingness of some officers and the inability of others to enforce discipline upon soldiers, there was at least an overall recognition that the wars ought to be fought within the context of a set of rules. This meant that when violence was inflicted upon civilians outside of those situations in which it was held to be permissible – in the context of a siege for example[51] – it was generally the result of

---

45 Quoted in ibid., p. 1149.

46 TT E113 (8) *Rules and Directions for the Government of Soldiers*, August 1642.

47 P. Tennant, 'People and Parish. South Warwickshire in the Civil War', in Richardson, *The English Civil Wars*, pp. 174, 164.

48 Duckenfield to Fairfax, 6 March 1643, in *Fairfax Correspondence*, vol I, p. 79; N. Wharton, 'Letters from a Subaltern Officer', *Archaeologia*, 35 (1853), pp. 310-34.

49 Slingsby, *Memoirs*, pp. 322; *Memoirs of Captain John Hodgson*, p. 137.

50 Quoted in Kenyon, *The Civil Wars*, p. 124.

51 It is worth noting that the massacre of civilians at the siege of Bolton referred to above was an officially sanctioned reprisal ordered by Prince Rupert for the killing of a royalist soldier.

unorganized, random outrages by soldiers who were out of control. Violence against civilians was not an act of deliberate policy.

A further reason for the relatively restricted level of violence against civilians was the realization that when the fighting stopped, then people would have to live together. Thus general Sir William Waller told combatants 'still to look upon one another as enemies that might live to be friends'; and for the same reason William Sedgewick urged that the 'sword might be dipped in oil rather than in blood.'[52] Perhaps paradoxically then, the fact that these were *civil* wars served, in the case of England, as a restraining influence. This contrasts, of course, with the common experience of other states undergoing civil conflict, such as France in the sixteenth century or Yugoslavia in the twentieth.[53] The difference lies in the fact that there were no major religious, ethnic, or cultural fault lines running between the combatants in seventeenth-century England, making it feasible to follow the advice of Waller and Sedgewick. To be sure, historians have increasingly stressed the extent to which the wars originated from religious disputes, and the Parliamentarian forces, especially the New Model Army formed in April 1645, had a pronounced religious character.[54] Puritan religious zeal led to destruction of church interiors and a ban on popular festivals including Christmas. Yet the violence engendered by religious disputes never reached the levels seen elsewhere. This was because religious disagreements in England were about matters of emphasis within the same Protestant outlook; the conflict was not between two groups where each regarded the other as heretical. The religious zeal of the New Model was undoubted, but it led to an increased emphasis upon military discipline: swearing, drunkenness, fornication and casual violence towards civilians, which might have been tolerated in other armies, was frowned upon. And if the zealots who committed acts of religious iconoclasm found the continued attachment of ordinary people to their old ways baffling, it did not seem worth killing them for this.

Moreover, if religion was a divisive issue in 1642, though not one to kill for, there was much that united the combatants in the English wars. Though the use of the Welsh and the Cornish language persisted, English provided a *lingua franca*; the combatants generally came from the same racial stock; and there were shared economic and social links. There were, of course, 'foreign' soldiers involved in the conflicts on English soil during the 1640s, who for reasons of culture, language, race or religion, were marked out as outsiders, and while not always clear-cut, there is some evidence to suggest a greater use of violence in incidents where they were involved. Thus, the bloody fury of the assault troops who massacred civilians at

---

52 Quoted in Donagan, 'Codes and Conduct', p. 74; her 'Atrocity, War Crime and Treason', p. 1166.

53 D. Crouzet, *Les guerriers de Dieu. La violence au temps des troubles de religion* (2 vols, Seyssel, 1990).

54 I. Gentles, *The New Model Army in England, Ireland and Scotland, 1645-1653* (Oxford, 1992), ch. 4; Wharton, 'Letters from a Subaltern Officer', pp. 310-34 for zealotry in Essex's army.

Leicester in 1645 may have been related to the fact that they were from mid-Wales, though it also owed something to the loss of 30 of their colleagues in the breach. Colonel Copley hinted that at another siege, his troops had slaughtered defenders because they were 'mostly Walloons'.[55] It was certainly the case during the first civil war that relations between the Scots forces and the civilians on whom they were billeted were frequently strained. General Leslie recognized the dangers and enjoined his soldiers to take 'all possible care to prevent all differences betwixt themselves and the subjects of England' so as to preserve 'the happy union betwixt the nations.'[56] It proved a vain hope, for the record of the Scots billeted in Yorkshire was a dismal one indeed, and their lengthy roster of offences included murder, rape, robbery, theft, and assault. 'None dare refuse to provide anything for their appetites … they kill us in hot blood, beat us in cold,' wrote one inhabitant; 'the over burdened country groanes under their oppressions,' claimed another.[57] We should be cautious before accepting such complaints at face value, yet there is a wealth of detail in the petitions against the Scots which transcends the prejudiced and formulaic quality of most such complaints, and suggests they were soundly based in fact.[58] Leslie had acknowledged that his army was always going to be difficult to control. Yet despite his assertion that he had always 'beene ready to heare the just complaints of the meanest, and to give them satisfaction and reparation,' he actually made matters worse, for he connived at protecting 'his men', ensuring that they avoided trial or punishment for their misdeeds by an English court.[59]

Aside from the Scots, the other significant 'foreign' element fighting on English soil was Irish. Again, there are indications that their behaviour towards civilians could be unusually brutal. For example, the Irish contingent in Colonel Washington's forces at Worcester gained a nasty reputation for their depredations in the surrounding area.[60] Yet overall, the record of Irish troops in England is remarkably free of atrocities against civilians, not least because the Irish troops were dispersed throughout the royalist forces. Indeed, the reverse was the case, for it was the Irish in England who suffered at the hands of the native population. English fear and loathing of the Catholic Irish soldiers led Parliament to exempt them from the protection of the rules of war: in October 1644 it ordered that no quarter should be given to 'any Irishman or to any papist born in Ireland, which shall be taken in arms against the Parliament

---

55   Copley to Fairfax, 6 March 1643, *Fairfax Correspondence*, vol I, p. 82.

56   Ordinance of 4 November 1644, quoted in C.S. Terry, *The Life and Campaigns of Alexander Leslie, First Earl of Leven* (London, 1899), p. 342, fte 1. This made detailed arrangements for billeting troops, amplifying the more general instructions contained in TT E30 (17) *Articles and Ordinances of Warre: For the Scots Army going into England*, 1643, and TT E30 (12) *Declaration of the Committees of Both Kingdoms for Billeting the Souldiers*, 1644.

57   TT E358 (18) *Declaration Concerning the Miserable Sufferings of the Countrie under some of the Scots Forces*; E362 (4), *A Continuation of Papers from the Scots Quarters*.

58   See especially TT E365 (9) *A Remonstrance Concerning the Misdemeanours of some of the Scots Souldiers in the County of Yorke*, December 1646.

59   Quoted in Terry, *The Life and Campaigns of Alexander Leslie*, p. 380.

60   Styles, 'Worcester During the Civil Wars', p. 211.

in England.'[61] And in an especially nasty incident following the battle of Naseby, troopers from the victorious Parliamentarian army came across a group of women camp followers who called out to them in a foreign tongue. The troopers killed or slashed the faces of some 300-400, believing them to be Irish.[62]

There is, then, some evidence to suggest that cultural, religious or racial differences between 'foreign' soldiers fighting on English soil and the civilian population could worsen military behaviour. Yet it is important not to exaggerate. The record of these outsiders was not blemished by the kind of atrocities which occurred elsewhere in the British Isles. And this should not surprise us. These 'foreign' soldiers comprised a small proportion of the armies; the Scots, who made up the largest contingent, were allies; and the Irish, who were the most distinct of these outsiders, were dispersed among the royalist forces.

Ireland and the Barbarization of Conflict: The Moral Economy of War, Indiscipline and the Unrestrained Use of Violence

If England represented the low point on the barometer of violence, Ireland represented the highest, both in respect of the scale and the nature of the violence. Thousands of civilians were attacked, violated and murdered by soldiers, and atrocity became routine. In large measure, this was because the 'moral economy' of war in Ireland stood in marked contrast to that in England. The moral climate in which the fighting took place was predicated firstly upon pre-existing attitudes towards the Irish. As a result of substantial immigration, perhaps 1 in 15 of the population in some parts of Ireland consisted of Protestant settlers from England and Scotland by 1640.[63] They wished to reform the 'backward' religion, culture and economy of the Catholic Irish. The settlers' view of the native Irish was shared by the population of England and most Lowland Scots. Barnaby Rich described them as 'more uncivil, more uncleanly, more barbarous and more brutish in their customs and demeanours than any other people in the known world'; while Walter Meredith wrote of Queen Elizabeth I 'wrastling in Ireland, to civilize and suppress these notorious and barbarous rebels.'[64] The treatment of civilians in the wars would be conditioned by this received opinion of the Irish as barbarous and uncivilized.

---

61 *Acts and Ordinances of the Interregnum, 1642-1660*, ed. C.H. Firth and R.S. Rait (3 vols, London, 1911), vol I, pp. 554-5.

62 In fact, the women were probably Welsh and had spoken in Gaelic Welsh. See P. Young, *Naseby 1645* (London, 1985), ch. 15 and Gardiner, *History of the Great Civil War*, vol III, p. 252 for the often conflicting primary sources.

63 There had been at least 100,000 immigrants between 1603 and 1641: N. Canny, 'What Really Happened in Ireland in 1641?', in Ohlmeyer, *Ireland from Independence to Occupation*, p.42; his *Kingdom and Colony; Ireland in the Atlantic World* (Baltimore, 1988), passim.

64 *A Short Survey of Ireland* (London, 1609), p. 2; TT E109 (8) *The Fidelity, Obedience and Valour of the English Nation*, August 1642. Another good example of England's 'civilizing mission' is in TT E44 (13) *The Impudence of the Roman Whore*, 1644.

It was even more powerfully shaped by what was *believed* to have happened when Ireland rebelled in 1641. There were reports from survivors, collected together and published in England, that 200,000-300,000 settlers had been killed, with men slaughtered, babies ripped from their mother's womb, and children roasted alive.[65] The author of *An Alarum to Warre* was typical in his blending together of atrocity tales. 'Those monsters of nature ... have not only sacked and pillaged wherever they have come, but they have taken delight to torture the Protestants without any bowels of compassion. They have ravished and deflowered chaste women and virgins, throwne small children into rivers with pitchforks; they stripped 1,500 stark naked ... drowned and killed many of them by the way. I tremble to speak of it, they have searcht women's privities for money, and in a most horrid and stupendious manner they have ript up their wombs great with childe, and slaine with the sword both them and their infants.'[66] One particular feature of the atrocities was the treatment accorded to women. Though the documented number of rapes remained 'curiously small',[67] there were manifold examples of women, from all social classes, being stripped of their clothing and physically abused, an action which may have been regarded contemporaneously as worse than rape. Stripping sometimes occurred in the context of a search for money, as *An Alarum to Warre* suggested, but generally it seems to have been designed to humiliate and degrade the victim and their families and to sever any future links between the parties concerned who might have been neighbours in the past. The precise extent of the atrocities remains a matter of dispute. Some outrages did undoubtedly occur, though the depositions of the alleged victims enormously exaggerated their scale. What was significant at the time, however, was that the atrocity stories were *credited* as being true. It is hard to exaggerate their impact upon public opinion outside Ireland. As Richard Baxter noted, 'There was nothing that wrought so much as the Irish massacre and rebellion.'[68]

The circumstances in which the rebellion began would largely determine the way in which the subsequent war would be fought. The atrocities generated a wave of revulsion, and they reinforced long-standing opinions about the barbarity of the Irish. It was but a short step to argue that therefore the Irish did not deserve the rights accorded to civilized people. In particular, they had forfeited the protection of the rules of war which gave protection to non-combatants. Irish 'blood-guilt' in 1641 thus provided a justification for the counter-atrocities that would subsequently

---

65 200,000 was 'the common report': *Memoirs of Captain John Hodgson*, p. 91. On the use to be made of the survivors' depositions as primary sources for the historian see M. Perceval-Maxwell, 'The Ulster Rising of 1641 and the Depositions', *Irish Historical Studies*, 21 (1978), pp. 144-67; and above all Canny 'What Really Happened in Ireland?', pp. 24-42. This recapitulates the excellent chapter on historiography from his *Making Ireland British, 1580-1650* (Oxford, 2003), which also provides a supremely balanced view of events.

66 TT E142 (6), 1642. See also TT E109 (16), *The Truest Intelligence from the Province of Munster*, 6 August 1642.

67 M. O'Dowd, 'Women and War in Ireland in the 1640s', in M. MacCurtain and M. O'Dowd (eds), *Women in Early Modern Ireland* (Edinburgh, 1991), p. 101.

68 *Reliquiae Baxterianae*, ed. M. Sylvester (London), 1696), part I, p. 26.

be committed by the Scots and English.[69] The atrocities also meant that William Waller's cautionary words, about the need for old neighbours to live as friends once the conflict was ended, could have no place in the Irish context. If civil war in England was a reason for the restraint of military violence, rebellion in Ireland instead provided a justification for bloody reprisals.

The Irish uprising additionally played to long-standing fears of English and Scots Protestants concerning Catholicism. It was not just that cruelties were to be expected of Catholics.[70] Just as significantly, the rebellion raised concerns that it was part of a wider, popish plot by England's enemies to obtain a 'back-doore' entry into the kingdom.[71] The rebellion was 'a sinister cause of introducing forraigne enemies', wrote Thomas Morton, who linked the Irish uprising to the troubles besetting France, Spain and Castile in 1642.[72] A similar theme was developed in *A New Discovery of the Design of the French*;[73] while another observer commented that, 'The contrivers of this Rebellion began to act first in Ireland, yet their aymes and purposes were, when that was subdued, to powre in great forces from thence into England.'[74] The Irish, then, were doubly culpable. They bore the 'blood-guilt' of 1641; and they were rebels who sought to introduce foreign enemies into England. They were therefore doubly undeserving of the protections normally accorded to enemies.

Against this background, it is unsurprising that the Scots and English soldiers acted towards Irish civilians with an indifferent and almost casual brutality, though the scale of this still retains the power to shock. Typical was an English soldier's account of events at Clannakeltie in the summer of 1642. 'Found in the town not above twenty men, women and children, which our troopers killed all, and ranged about, and found some hundred more hid in gardens and killed all; there you might have seen every sex discovered, and some lying on their backs, old, young, none

---

69 See, for example, the justification of the slaughter of garrison troops and civilians at Drogheda and Wexford by Cromwell's men: *Oliver Cromwell's Letters and Speeches*, ed. T. Carlyle (London, nd.), pp. 298, 311.

70 Thus Thomas Morton compared the cruelties of 'the most wicked papists' in Ireland to that of the Turks, 'a people so cruell that they make it a point of religion and a work meritorious to murder God's people': TT E109 (14), *Thos Morton. England's Warning Piece, Shewing the Nature, Danger and Ill Effects of Civill-Warre*, 5 August 1642, p. 3.

71 Comment from House of Lords, quoted in K.J. Lindley, 'The Impact of the 1641 Rebellion upon England and Wales, 1641-5', *Irish Historical Studies*, 18 (1972), p. 160. Fears of popish conspiracy were not new. Cornwallis, the English ambassador to Spain had earlier warned of the dangers from Irish troops who served with the armies of Spain, fearing that 'they may be hereafter enabled to further mischief': quoted in Casway, 'Henry O'Neill and the Formation of the Irish Regiment in the Netherlands, 1605', p. 485. See also E. H. Shagan, 'Constructing Discord: Ideology, Propaganda and English Responses to the Irish Rebellion of 1641', *Journal of British Studies*, 36 (1997), p. 23.

72 *Thos Morton. England's Warning Piece*, p. 4.

73 TT E109 (21) 21 August 1642.

74 TT E138 (22) *A Relation Touching the Present State and Condition of Ireland*, February 1641.

spared.'[75] John Bacon too wrote in a similarly laconic vein: 'We took twentie of the rebels and hanged them up.'[76] In 1644, an English officer recorded: 'Our men burned the house, and killed a woman or two, marched on.'[77] As this brief and uncaring diary entry suggests, the neglect of the rules of war proceeded with official connivance, and in many instances the men were just following the example of their officers. Thus, at the surrender of Newry in May 1641, the killing of civilians was sanctioned by the commander of the Scots forces, Robert Monro, and it was Lord Conway, Marshal of Ireland, who was 'the principall actor' in the slaughter. 'Our sojors ... seeing such prankes playd by authoritie ... thought they might doe as much any where els,' wrote James Turner, a seasoned campaigner from the German wars.[78] Official sanction for the military violence came from the highest levels. As the Lords Justices explained, 'We have ... proceeded against the rebels with fire and sword, the soldiers not sparing the women and sometimes not children, many women being manifestly very deep in the guilt of this rebellion.'[79] So routine was the brutality of the soldiers, and such was the extent of official approval for their actions, that it is legitimate to refer to the use of violence against civilians as an act of policy.

Two additional factors compounded the soldiers' propensity to use violence against civilians. The first was the indiscipline of many of the troops. The atrocities of 1641 were themselves partly the work of disorderly soldiers as well as an 'angry peasantry'.[80] The insurgents at the outset of the Irish rebellion included reliable troops from Wentworth's Catholic army of 1640[81] and recruits for Spanish service, on the one hand, alongside peasants and *creaghts*,[82] on the other. Despite the presence of regular troops, this motley collection of forces was poorly-controlled by its commander, Sir Phelim O'Neill, who proved unable to prevent the commission of atrocities. Subsequently, the arrival from the Netherlands of the highly respected Owen Roe O'Neill and Thomas Preston, who took over command of the Armies of Ulster and Leinster respectively, plus the expertise of returning veterans, saw some improvement in discipline, though even O'Neill found it next to impossible to work with the *creaghts*.[83] The Irish Confederate forces always consisted of loose coalitions and it was never easy to control them. Similarly, discipline among Monro's Scots

---

75 TT, E109 *Good Newes from Ireland: From these Severall Places, namely Kimsale, Bandum, Clarakelty*, 4 August 1642, p. 3.

76 TT E138 (2) *Letter from John Bacon under Sir Simon Harcourt*, 12 February 1642.

77 Quoted in Carlton, 'The Face of Battle', pp. 243-4.

78 *Memoirs of his own Life and Times* (Edinburgh, 1829), p. 24.

79 Quoted in O'Dowd, 'Women and War', p. 100.

80 Perceval-Maxwell, 'The Ulster Rising of 1641', p. 165.

81 When rebellion had erupted in Scotland in 1640, the king's deputy in Ireland, Sir Thomas Wentworth, had assembled an army, which in fact never left Ireland.

82 Nomadic Irish cattle herders who accompanied the armies and contributed to their victualling.

83 S. Wheeler, 'Four Armies in Ireland', in Ohlmeyer, *Ireland from Independence to Occupation*, p. 47. On O'Neill see also J.I. Casway, *Owen Roe O'Neill and the Struggle for Catholic Ireland* (Philadelphia, 1984).

forces, which may have numbered 10,000 by the spring of 1642, was not good. The troops were initially contracted to be paid by the English Parliament, but when civil war broke out in England, it became clear that no money would be forthcoming. Discipline broke down, and widespread plundering and violence occurred, with the officers frequently in the lead, as Sir James Turner's *Memoirs* make clear.[84]

The one exception to this sorry tale of military ill-discipline was Cromwell's army of conquest of 1649-50.[85] A section of the New Model objected to the Irish campaign and had to be weeded out of the expeditionary force in Summer 1649. Amongst them were some members of the Leveller tendency who objected on the principled grounds that all men had equal rights and it was wrong to impose a tyranny on the Irish by conquest, though the evidence for their views is not clear-cut.[86] The majority of soldiers certainly did not share the Levellers' scruples, yet once in Ireland they behaved with notable restraint towards civilians. The reason for this was not their novel views of the Irish, but rather the discipline that Cromwell was able to impose on the army. Supremely well supported by the Commonwealth Government, which supplied food, clothes and money, he was able to enforce a harsh authority upon his men. To be sure, he acted ruthlessly against rebels captured in arms, and slaughtered garrisons and civilians in Drogheda and Wexford in order to induce towns to capitulate subsequently. Yet outside of these circumstances, the threat of punishment ensured that there was little random violence by his soldiers towards civilians. Cromwell had no love for the Irish, but he recognized that severe discipline was necessary to preserve the efficiency of the army as a fighting force.[87] The discipline of Cromwell's forces stands out by its rarity, however.

The lax discipline of the majority of forces in Ireland thus resulted in barbaric treatment of civilians. However, even more important in this regard was the nature of the warfare conducted by the armies. First, all sides made great use of what might loosely be termed scorched-earth tactics. The Confederate forces rarely offered battle. Instead, they resorted to ambushes, raids and the destruction of buildings and supplies. 'The whole design of the rebels we saw was to starve us, by burning all the corne and hay within two miles of us,' noted one of the English commanders.[88] Lord Clanmaliry, one of the rebels, agreed that Irish tactics were 'to weary them [their

---

84 The army 'fingered no pay the whole time I stayd in Ireland, except for three months': Turner, *Memoirs*, p. 24.

85 For lack of pay and problems of discipline among English forces earlier, see for example TT E134 (26) *A True Copy of a Letter from Dublin*, 8 February 1642; E86 (38), *Special Passages*, January 1642.

86 C. Durston, '"Let Ireland Be Quiet": Opposition in England to the Cromwellian Conquest of Ireland', *History Workshop Journal*, 21 (1986), pp. 105-12; R. Foxley, *Citizenship and the English Nation in Leveller Thought*, Unpublished PhD thesis (Cambridge, 2001), pp. 102-7.

87 J.S. Wheeler, 'Logistics and Supply in Cromwell's Conquest of Ireland', in Fissel (ed.), *War and Government*, pp. 38-56.

88 TT E140 (2) *A Letter Sent for Sir Simon Harcourt*, 18 March 1641.

opponents] out by our surprises and depredations.'[89] In retaliation, the Protestant forces, led by Monro in Ulster, Ormond in Dublin and Inchiquin in Munster, responded in kind, destroying crops and buildings and taking reprisals on civilians accused of sheltering the rebels. Monro left Armagh and Charlemont wasted and with 'a wonderfully deserted appearance.'[90] Colonel Crafford was reported 'to have done good service by burning above 200 townes' in County Wicklow alone, his soldiers 'killing what they found out of castles, which were many scattering rogues.'[91] Another commander, Henry Tichbourne, 'burnt severall townes about him for three or foure miles ... The other side are made to see that now they have don (sic) pillaging as we begin to pillage them.'[92] Second, the Confederate forces generally did not wear uniform, making it hard for their opponents to distinguish soldiers from civilians, though in practice little attempt was made, so contemptuous of the Irish as a whole were the Protestant forces.[93] Edmund Ludlow remembered coming across an encampment containing half a dozen families. 'Some of those [soldiers] who saw them first, *presuming all the Irish in that country to be enemies*, began to kill them; of which, having notice, I put a stop to it,' he wrote.[94] Significantly, he did not take any action to punish the perpetrators of the violence. In short, then, there was little differentiation between combatants and non-combatants in the fighting in Ireland.

Scotland: Complexities of Race and Culture and the Institutionalization of Violence

If we turn finally, and briefly, to consider the warfare in the last of the three kingdoms, we find that Scotland occupies a mid-point between England and Ireland on the barometer of military violence. The scale of the violence, in terms of the numbers of people affected, was less extensive than in Ireland, largely because the number of soldiers involved in the conflicts was not as great. The armies were typically no more than 2,000-3,000 strong. Even at the largest of the battles, Kilsyth in 1645, the total number on both sides was no more than 10,000-12,000. Equally, the size of the garrison establishment was relatively small. By contrast, there were 27,000-35,000

---

89 Ludlow, *Memoirs*, vol I, p. 240.

90 Casway, *Owen Roe O'Neill*, p. 111. On Ormond see p. 159.

91 TT E108 (46) *A True Relation of Such Passages and Proceedings of the Army of Dublin*, 3 August 1642, pp. 3-4.

92 TT E141 (26) *A Certaine Relation of the Earl of Ormonde's Nine Days Passages*, March 1642.

93 On the lack of uniform, see TT E114 (26) *Exceeding Joyful News from Ireland*, 27 August 1642; E108 (47) *A True Relation of the Taking of Mountjoy in the County of Tyrone by Colonel Clotworthy*, 4 August 1642 with examples of how the English used this to their advantage.

94 Ludlow, *Memoirs*, vol I, p. 270, my italics. For another example, see TT E138 (2) *Letter from John Bacon under Sir Simon Harcourt*, 12 February 1642.

Protestant and 14,000-22,000 Catholic troops under arms in Ireland by 1643.[95] Yet if the scale of the violence was dissimilar to that in Ireland, the nature of it was not. For the warfare in Scotland was also characterized by cruelty, disregard for the welfare of civilians and atrocities. A partial explanation for this lies in the composition of the forces on both sides. They comprised loose coalitions rather than regular armies, and were not amenable to discipline.[96] The problems of discipline were made worse by the fact that many recruits had no previous experience of soldiering – one observer described them as 'a mere rabble' – and were merely interested in thieving.[97] However, a more important explanation is to be found in the racial and cultural tensions which were woven into the conflicts.

The fault lines which divided one group from another in Scotland were complex.[98] On the one hand, there was a division between Highlanders and Lowlanders. The former were Catholic for the most part, spoke Gaelic, operated a pastoral economy, wore distinctive dress, were indifferent to the claims of central government and feuded as a way of life. By contrast, large sections of the population of Lowland Scotland had become Calvinist after the Reformation, spoke Scots or English, aligned themselves with the Protestant English *Gaill* who had settled in the Lowlands, had a more commercially oriented economy and a more settled way of life. To be sure, it is important not to exaggerate the differences. Lowland Scotland was also lawless and feuding; its economy was also agricultural; and it retained pockets of Catholicism. Nevertheless, contemporaries were aware of differences, even if with hindsight they appear less stark than at the time. Lowlanders certainly regarded their neighbours as backward and uncivilized, their hankerings after Catholicism as deeply suspicious and wished to reform the Highlanders' language, economy and social structure. For their part, many Highlanders regarded themselves as still having more in common with the Gaelic Irish than with their Lowland neighbours. Accordingly, it is not unfair to think in terms of an incipient 'cultural war' between Highlanders and Lowlanders in the seventeenth century.

Confusingly, these divisions between Highland and Lowland were overlain by clan or tribal loyalties. What might be called the clan structure – though this term implies an overarching uniformity and immutability to clan loyalties which did not

---

95 D. Stevenson, *Alasdair MacColla and the Highland Problem in the 17th Century* (Edinburgh, 1980), pp. 142, 201; Ohlmeyer, *Ireland from Independence to Occupation*, p. xxiii.

96 S. Reid, *The Campaigns of Montrose. A Military History of the Civil War in Scotland, 1639 to 1646* (Edinburgh, 1990), esp. pp. 155, 168-9.

97 G. Wishart, *The Memoirs of James, Marquis of Montrose* (London, 1903), pp. 82-3.

98 For much of what follows see C.J. Withers, *Gaelic Scotland: The Transformation of a Cultural Region*; D. Stevenson, 'The Century of the Three Kingdoms', in J. Wormald (ed.), *Scotland Revisited* (London, 1991); A.I. Macinnes, 'Gaelic Culture in the Seventeenth Century: Polarization and Assimilation', in S.G. Ellis & S. Barber (eds), *Conquest and Union. Fashioning the British State, 1485-1725* (Harlow, 1995), pp. 162-94; S.G. Ellis, 'The Collapse of the Gaelic World, 1450-1650', *Irish Historical Studies*, 31 (1999), p. 452 who notes that those who spoke Gaelic might well be referred to as 'Irish' by speakers of English or Scots.

really exist – did not altogether square with the Highland–Lowland division. The most powerful of the clans by the first half of the seventeenth century were the Campbells whose lands lay in the Highlands, though they associated themselves with Lowland aspirations and detestation of the Highlanders. Accordingly, the alliances which emerged in the course of the Scottish conflict are not easy to describe. There was a war of Highlanders and Irish Gaels on the one hand, against Lowlanders on the other. Yet the leader of the former, the royalist Marquis of Montrose, was a Lowlander; not all Highlanders were Catholic; and many thought the real war they were fighting was against Campbell hegemony.

However, if the outlines of the alliances are not always clear cut, at least to modern eyes, there is no denying the very real cultural, racial, religious and linguistic tensions which embittered the fighting between the soldiers and produced atrocities against civilians. One of the best known examples occurred after the battle of Lagganmore. Defeated Campbell soldiers together with women and children from the area were herded into a barn by the victorious troops, many of whom belonged to the rival Donald clan. On the orders of their commander, Alasdair MacColla, the barn was fired. All perished, except for a single woman and John Campbell, one of the defeated captains.[99] The Lagganmore atrocity occurred in the course of an attack upon Campbell lands in 1645-46. Such attacks went beyond mere plundering for the sake of booty. Instead, they involved the deliberate destruction of crops, property and the killing of civilians in a process euphemistically referred to as 'wasting'. It was in this context that much of the military violence against civilians took place. 'Wasting' was a distinctive element of Scottish warfare and sprang from the clan and cultural tensions which informed so much of the fighting.[100] Harrying an opponent's lands was a traditional aspect of inter-clan warfare, but the practice seems to have gained in scale and intensity after 1643 as the clan feuding gave way to civil war within Scotland.

'Wasting' served a number of purposes. It provided booty, the cement that held the armies together and without which they dissolved. Thus many of the men from the clan Donald and MacLean quit Montrose's army in disgust after the huge victory at Kilsyth because their commander would not permit the plunder of Glasgow.[101] 'Wasting' was also a means of destroying a rival's power. This was the apparent motive behind the attack upon the lands of the Campbells in Argyll in 1644-45 and on a second occasion in the following year. MacColla claimed that he 'left neither house nor hold unburned, nor corn nor cattle, that belonged to the whole name of Campbell'; while some 895 young men of military age were killed in order to fetter Campbell military capacity in the future.[102] 'Wasting' was additionally employed against an enemy's allies. MacColla's men attacked the lands of the Frasers and

---

99 Stevenson, *Alasdair MacColla*, pp. 218-20.

100 K.M. Brown, *Bloodfeud in Scotland: violence, Justice and Politics in an Early Modern Society* (Edinburgh, 1986).

101 Stevenson, *Alasdair MacColla*, pp. 204-7.

102 Stevenson, *Alasdair MacColla*, pp. 147-48, 213-7.

others in eastern Ross-shire for having fought at Auldearn, his men behaving like 'fiends' as they looted and raped. Unsurprisingly, the Campbells retaliated in kind against allies of Montrose. It was also useful as a means of coercing neutrals, for many clans stood aside, waiting to see which way the conflict would go before committing themselves.[103] Yet whatever the particular reasons behind 'wasting', it was legitimated in the eyes of its practitioners by cultural differences and clan rivalries, and civilians were seen as fair game. In effect, the practice institutionalized the use of military violence against civilians.

The presence of Irish soldiers on Scottish soil further complicated the picture. In 1644, Irish troops had been sent to Scotland by the Earl of Antrim, the rebel leader in Ireland, to fight alongside the royalist forces. The hope was that their presence would force the withdrawal of the Scottish army from Ireland, though this hope proved vain. The Irish contingent in the royalist forces was substantial and played a major part in the fighting and in the pillaging of civilians, for which they gained a particularly unsavoury reputation. Patrick Gordon commented on the offhand way in which they killed and looted even poor labourers.[104] They ransacked Aberdeen, 'hewing and cutting down all manner of man they could avertak within the toune.' Victims were made to undress before they were killed so that their clothes would not be stained with blood. The plundering and raping of women continued from Friday until Monday, then large numbers of women were taken back to camp to be ravished at leisure.[105] Such reports might be dismissed as propaganda, except that they come from eyewitnesses, and ones moreover who were sympathetic to the royalist cause.

In some ways, the Irish were simply importing into Scottish warfare the kinds of brutalities which had been practiced against them by both English and Scots forces in their homeland. There were old scores to settle, and the Irish settled them in full. For them, the war also had a racial side to it; as part of the broader Gaelic community they slotted into Scottish civil wars which pitted Highlander against Lowlander and Gael against Gael. Importantly, however, they probably came to feel that they had little to lose. Unlike the Highlanders alongside whom they fought, the Irish could not drift back to their homes at the conclusion of a campaign, and plunder was the only means they had to sustain themselves. Moreover, every man's hand was turned against them – even Montrose was obliged to take severe action against them in 1646[106] – and from their point of view there was little to be gained by showing mercy to the local population.

The brutal treatment of civilians by the Irish soldiers in Scotland contrasted markedly with the behaviour of their fellow-countrymen attached to the royalist

103   Stevenson, *Alasdair MacColla*, p. 126 on Sir Alexander Menzies whose lands were wasted so that 'on the threshold of war he might terrify others'.

104   *A Short Abridgement of Britane's Distemper*, ed. J. Dunn (Aberdeen, 1844), pp. 160-61.

105   J. Spalding, *Memorials of the Troubles in Scotland* (2 vols, Aberdeen, 1850-51), quoted in Stevenson, *Alasdair MacColla*, p. 135; Bennett, *The Civil Wars*, pp. 238-9.

106   M. Napier (ed.), *Memorials of Montrose and his times* (2 vols, Glasgow, 1848), vol II, pp. 204-6.

forces in England, and also with the conduct of the English forces under Cromwell who conquered Scotland in 1650-51. Although, as noted earlier, a shortage of supplies led to some ill-disciplined behaviour in the summer of 1650, overall these soldiers were reasonably well-behaved towards civilians. A partial explanation for this may again be found in the famous discipline of the New Model Army as well as in the nature of the fighting. After the decisive English victory at the battle of Dunbar, the Scots forces did not fracture into guerrilla units, and the Scots never took up arms as a nation. Consequently, the English army was able to prosecute the campaign as a conventional conflict with all the traditional restraints on the use of violence. However, an additional and possibly more significant reason for the relative absence of barbarism in this phase of the conflicts was the English soldiers' view of the Scots. As James Wheeler notes, the English army and the Scots had a 'common religion and a shared history of resistance to Stuart tyranny in the 1640s' before the latter were led astray by the perfidy of Charles II.[107] Consequently, it was possible to regard the Scots as deluded 'Brethren'. Indeed, this was how Cromwell and the General Council of the Army referred to them.[108]

Violence and a Sense of the 'Other'

Even a brief survey of the conflicts in the British Isles during the 1640s reveals, then, that the incidence and nature of military violence against civilians could vary quite significantly. And this is not surprising for, as we have also seen, the factors which conditioned soldiers' behaviour were both varied and complex. They included the military traditions of the combatants, the availability of pay and supplies, the ability and willingness of officers to impose discipline, the awareness of codes of conduct and the readiness to apply them, and the political circumstances in which the wars were fought, and the attitudes of the troops' political masters. What perhaps emerges above all, however, is that soldiers' behaviour towards civilians was crucially conditioned by their sense of the 'Other'.[109] Where there were significant racial, religious or cultural differences between soldiers and civilians, then these might be sufficient to override the constraints which otherwise restrained the use of military violence. This, surely, is a theme which resonates through the centuries. Mutual cultural incomprehension between Welsh and English in the twelfth and thirteenth centuries allowed the commission of acts of extreme cruelty.[110] The same

---

107   J. S. Wheeler, 'Sense of Identity in the Army of the English Republic, 1645-51', in A.I. Macinnes and J. Ohlmeyer (eds), *The Stuart Kingdoms in the Seventeenth Century* (Dublin, 2002), p. 163.

108   Wheeler, 'Sense of identity', pp. 162-3.

109   V. Harle, 'On the Concept of the "Other" and the "Enemy"', *History of European Ideas*, 19 (1994), pp. 27-34, suggests that the sense of Other develops when a group or individual displays characteristics which are alien to one's own self image.

110   F. Suppe, 'The Cultural Significance of Decapitation in High Medieval Wales and the Marches', *The Bulletin of the Board of Celtic Studies* 36 (1989), pp. 147-60. See also M.

was true of the Albigensian Crusades which, as Malcolm Barber has noted, 'went far beyond the normal conventions of early thirteenth-century warfare, in the scale of the slaughter, in the execution of high-status opponents, male and female, in the mutilation of prisoners, in the shaming of the defeated, and in the quite overt use of terror as a method of achieving one's goals.'[111] It is, of course, important to make such comparisons across the centuries with care. The English and Scots certainly regarded the Irish as backward and uncivilized. The Scots Lowlanders regarded their Highland neighbours in the same light. But they did not regard these inferiors as subhuman. And this perhaps prevented the commission of the special kinds of atrocities, including ethnic cleansing, practiced in the twentieth century by the *Waffen SS* and the Serbs in Eastern Europe. The record of soldiers in the 1640s was blemished: yet their behaviour did not plumb the depths of awfulness witnessed in subsequent centuries.

Strickland, *War and Chivalry. The Conduct and Perception of War in England and Normandy, 1066-1217* (Cambridge, 1996).

111   'The Albigensian Crusades: Wars like any Other?', in M. Balard, B.Z. Kedar and J. Riley-Smith (eds), *Gesta Dei Per Francos. Crusade Studies in Honour of Jean Richard* (Aldershot, 2001), pp. 45-55. I am grateful to Professor Barber for drawing my attention to the above works by Suppe and Strickland.

Chapter 3

Hatred and Honour in the Military Culture of the French Revolution

Ian Germani

The French Revolution is conventionally represented as a watershed in the escalation of warfare. The *levée en masse* of 23 August 1793, with its call for the mobilization of every French man, woman and child, represented the ideal of the nation-in-arms and, at least in inception, the concept of total war. Clausewitz himself articulated this idea: 'War, untrammelled by any conventional restraints,' he wrote, 'had broken loose in all its elemental fury.'[1] By infusing warfare with new ideological significance, the revolutionaries had abandoned the conventions that at least aspired to control and limit its violence during the age of enlightenment. Robespierre himself starkly outlined this rupture with the past as he justified the National Convention's decree that no English or Hanoverian troops were to be taken prisoner. 'What is there in common between what existed before and what now is?' he asked. 'What is there in common between liberty and despotism, between crime and virtue?'[2] There could, he said, be no fellow feeling between the soldiers of liberty and the 'slaves of the tyrants' whom they fought. A few days earlier, in presenting the decree to the Convention, Bertrand Barère had insisted that French soldiers who had English troops at their mercy should remember:

> the vast countries that the English emissaries have devastated. Cast your view on the Vendée, Toulon, Landrecies, Martinique and Saint Domingue. These places are still soaked in the blood which the atrocious policy of the English caused to flow.

When victory provided the occasion, said Barère, the French soldier should strike, so that not a single English soldier survived: 'Let the English slaves perish, and Europe will be free.'[3]

The Jacobin discourse on war, articulated by Robespierre and Barère, insisted upon the revolutionary rupture with the past and the repudiation of traditional constraints, including the obligation to show clemency to a defeated foe. It also sought to inspire

---

1   Carl von Clausewitz, *On War*, ed. and trans. by Michael Howard and Peter Paret (Princeton, 1973), p. 593.

2   *Moniteur*, 24 June, 1794.

3   *Moniteur*, 29 May, 1794.

hatred for an enemy who, by virtue of his barbarous conduct, had placed himself beyond the pale of humanity. This discourse alone would seem to confirm that the Revolution contributed to barbarization of warfare. Yet most historians are quick to point out that the Convention's decree concerning English and Hanoverian troops went unheeded by republican forces. With some notable exceptions, the accepted rules of war were respected, at least insofar as regular combatants were concerned.[4] The purpose of this essay is to consider why this was so. Its argument is that the ideologically charged Jacobin discourse of hate was moderated by the Jacobins' own idealization of war as liberation as well as by their preoccupation with the internal enemy rather than the external one. French traitors and *émigrés* ranked higher in the orders of villainy than the 'slaves of the tyrants' who had invaded French soil. The Jacobin discourse was also counterbalanced by a more conservative discourse on war which elevated the soldier's professional code of honour above political or ideological considerations. Never completely distinct, these rival discourses waxed and waned in relation to one another according to the political vagaries of the Revolution itself.[5] Ultimately, however, the warrior's code of honour worked, in the absence of any more effective constraints, to limit the barbarity in warfare that was promised by the ideologically charged context of the Revolution. Particular attention to the Army of the Sambre and Meuse during its campaigns in Germany in 1795 and 1796 suggests that, even *in extremis*, the soldier's code provided a barrier, albeit a fragile one, against the barbarization of warfare.

The Jacobin Discourse of Hatred

The Jacobin discourse on war, which culminated during the Terror of 1793-94, was part of a broader discourse which has been exhaustively analyzed by historians.[6] One of the most remarkable aspects of that discourse was the immense effort dedicated to its propagation and the variety of media employed to that end.[7] Newspapers, plays, poems, visual images and civic festivals were all employed to inspire love of the fatherland and the determination 'to live free or die'. This multi-media propaganda

---

4   Geoffrey Best, *Humanity in Warfare* (New York, 1980), p. 81.

5   It is important to note that the revolutionaries themselves, in this respect as in so many others, exaggerated the rupture with the old regime. As David Bell has demonstrated, the Seven Years' War represented an important watershed in terms of the investment of warfare with nationalist ideology. Already, in that conflict, French propagandists represented English soldiers as 'barbarians'. David Bell, *The Cult of the Nation in France: Inventing Nationalism, 1680-1800* (Cambridge, Massachusetts, 2001), pp. 78-106.

6   In particular, see François Furet, *Interpreting the French Revolution*, trans. by Elborg Forster (Cambridge, 1981) and Mona Ozouf, 'War and Terror in French Revolutionary Discourse', in *The Journal of Modern History*, 56, 1984, pp. 579-597.

7   James A. Leith, *Media and Revolution: Moulding a New Citizenry in France during the Terror* (Toronto, 1968); Serge Bianchi, *La revolution culturelle de l'an II: Elites et people 1789-1799* (Paris, 1982); Jean-Claude Bonnet, ed., *La Carmagnole des Muses: l'Homme de Lettres et l'Artiste dans la Révolution* (Paris, 1988).

campaign was directed particularly at the army, which became, during the Terror, 'the school of Jacobinism'.[8] A significant example of this propaganda is Jacques-René Hébert's radical newspaper, *Le Père Duchesne*, the pages of which exalted the revolutionary war effort and excoriated the enemy in the ripe slang of the common man. Issue number 297, for example, expressed:

> The Great Fury of Père Duchesne against the infamous soldiers of Turkey-King George [*Georges Dandin*] who, after taking Toulon through treason, have hanged the representatives of the people, pillaged the homes and massacred women and children.[9]

Père Duchesne's advice to the 'brave sans-culottes' was 'to exterminate all the mad dogs unleashed by Pitt to ravage France and reduce us to slavery.' Hébert's language, justifying a war of extermination by reference to the enemy's inhumanity, was the same as that of Barère or Robespierre. Other media echoed the same refrain. An engraving depicted the English flight from Toulon in December, 1793, representing the English as cowardly and rapacious, more interested in carrying off booty and women than in fighting the French. The print made the point that this enemy deserved no quarter by representing French soldiers mercilessly butchering their foes with cold steel. Plays and festivals developed the same message. Again, to use the propaganda surrounding the siege of Toulon as an example, Briois's play, *La Prise de Toulon*, focused upon the villainous English officer, Volmer,[10] whose amorous advances to the female heroine, Millette, were honourably rebuffed. As an aristocrat undeserving of mercy, Volmer is denied the opportunity to surrender by Millette's lover, who dispatches him with his sword at the conclusion of the play.[11] Finally, festivals celebrating the recapture of Toulon were often iconoclastic revels marked by the symbolic degradation, in effigy, of William Pitt, the Prince of Cobourg and other enemy figures.

It was not just the English who were vilified in this way. At the outset of the war, in 1792, Austrian and Prussian troops were also charged in the republican press with the commission of atrocities. *L'Ami Jacques, Argus du Département du Nord*, reported on 29 May 1792 that Austrian forces, 'these veritable murderers and brigands', had massacred French wounded following an action near Liège and as they withdrew, 'they pillaged and committed all sorts of horrors.'[12] On 7 September, reporting on Austrian actions in the vicinity of Verdun, the journal stated, 'The uhlans committed horrors there which make one shudder; they sabred pregnant women and cut the heads off children, etc.' The paper noted with satisfaction that a local woman had exacted vengeance by poisoning two barrels of wine which had subsequently been consumed by four hundred enemy soldiers.[13] On 10 October, the journal went

---

8   Jean-Paul Bertaud, *La Révolution Armée: Les soldats-citoyens et la Révolution française* (Paris, 1979), pp. 194-229.

9   *Père Duchesne*, No. 297.

10   'Volmer' is a play on words – literally, 'sea-thief'.

11   Briois, *La Prise de Toulon, Drame en Deux Actes et en Prose* (Paris, 1794).

12   *L'Ami Jacques, Argus du Département du Nord*, 29 May, 1792.

13   Ibid., 7 September, 1792.

so far as to accuse the Austrian and *émigré* forces besieging Lille of crucifying Belgians and volunteers whom they had taken prisoner. The Archduchess Christine, the 'Austrian shrew', was accused of coming to witness 'this frightful theatre of barbarism and ferocity', encouraging the Austrian generals to fire heated shot into the city.[14] Some right-wing journalists angrily denied these charges.[15] Others, like Montjoye, more amusingly derided the revolutionaries' atrocity stories:

> The thousand times cruel Austrians, they say, are bringing with them six hundred anthropophagi, well chained and well muzzled, that they feed on twenty-five head of cattle each day, and that they intend to unleash upon us when they enter France. There are some peasants, who see clearly enough, who say the anthropophagi are to be found in France.[16]

Montjoye recognized that the hyperbole of revolutionary propagandists might be turned against them.

The propaganda denouncing enemy atrocities was far from all on one side. In July 1792, the right-wing press made much of the razing of a suburb of Courtrai by French troops under the command of General Jarry. The *Journal Général du Département du Pas-de-Calais* described how the 'Jacobins' had thrust an old man, an infant in his arms, back into the flames of his burning house; and how they had forced a child to jump to his death from a burning roof-top.[17] French soldiers' crimes against women were emphasized. 'They have quite recently committed many rapes,' stated the journal, concerning which their general 'is obliged to keep silent and to close his eyes.'[18]

There was an inevitable double-standard in the press treatment of military atrocity. The firing of heated shot against Lille by Austrians was described as a war crime in the republican press, while its employment against English vessels in the harbour at Toulon was considered a just reprisal. In part, however, the accusations and counter-accusations stemmed directly from the Revolution's new conception of warfare, in which every citizen was a potential combatant. In the opening days of the war, letters were exchanged between French and Austrian generals, in which they outlined their views concerning acceptable military practices. The Austrian Major-General Borros complained to General Luckner of attacks by French peasants on his troops and warned of legitimate reprisals in the event of further incidents. He wrote:

> I am convinced that the French want as much as we do to prevent the repetition of such incidents and that they will prefer to fight a fair and open war, in which the inhabitants of the countryside will have no active part.[19]

---

14 Ibid., 10 October, 1792.

15 *Journal Général du Département du Pas-de-Calais*, 23 October, 1792.

16 *Ami du Roi*, 8 August, 1792.

17 *Journal Général du Département du Pas-de-Calais*, 11 July, 1792.

18 Ibid., 1 July, 1792.

19 Archives de Guerre, B1-2: letter of 29 May, 1792.

Luckner's reply promised good conduct on the part of French troops, providing assurance of exemplary punishment of those who had committed excesses in massacring Austrian prisoners at Lille, but he refrained from promising to restrain civilians from defending their lives and property.[20] As he put it in a letter to the Minister of War, 'I refused to take it upon myself to accept his proposition and thereby to deprive France of an infinity of defenders.'[21] Similarly, General La Nouë, commandant of the French garrison at Maubeuge, responded to an Austrian complaint of civilian attacks by stating that the peasants' action should be considered 'a legitimate defence of their properties' against the threat of Austrian pillage.[22] Predictably, the French royalist press justified Austrian reprisals on both civilians and soldiers. The *Journal Général du Département du Pas-de-Calais*, justifying the bombardment of Lille, stated that, 'We consider both the Citizen-Soldiers and the Soldier-Citizens to be rebels' and therefore 'it is hardly surprising that the enemy fire is divided between the Soldier-Citizens and the Citizen-Soldiers.'[23]

Both sides, therefore, claimed that the severity of their soldiers' conduct was necessitated by the enemy's contravention of the laws of war. Both sides used stories of enemy atrocities to heighten the ideological significance of the war and to insist that, because of the principles at stake – freedom and equality on the one hand, social order and political authority on the other – the normal constraints no longer applied.

A War of Liberation

The violence of revolutionary rhetoric concerning the war was mitigated, however, by several factors. Firstly, while Jacobin publicists might justify merciless treatment of enemy soldiers, characterized as the 'slaves of tyrants', they emphatically denied any hostility to the people over whom the tyrants exercise their dominion. In this respect, revolutionary propaganda remained true to the idealistic aspirations of the National Assembly's so-called 'declaration of peace to the world' of 22 May 1790, which renounced the use of force against the freedom any people.[24] In fact, the revolutionaries insisted that their armies fought, first, to defend French liberty and, second, to extend its benefits to the oppressed peoples of Europe. In contemplating their foreign enemies, they made a distinction between those who benefited from the oppression of the old regime and those who did not. Thus, in the above-mentioned play by Briois on the taking of Toulon, the aristocratic Volmer is juxtaposed with an honourable English officer, Tumbridge, who predicts French victory.[25] A later

---

20 Archives de Guerre, B1-2: letter of 30 May, 1792.
21 Archives de Guerre, B1-2: letter of 31 May, 1792.
22 Archives de Guerre, B1-2: letter of 1 June, 1792.
23 *Journal Général du Département du Pas-de-Calais*, 3 October, 1792.
24 Jacques Godechot, *La Grande Nation: l'Expansion Révolutionnaire de la France dans le Monde de 1789 à 1799* (Paris, 1983), p. 66.
25 Briois, *La Prise de Toulon*.

play, Mittié's *Descente en Angleterre* of 1798, has French soldiers come to the rescue of English 'patriots' as the latter are about to be executed for conspiring to overthrow the monarchy.[26] In the press, the Committee of Public Safety sponsored journal, the *Anti-fédéraliste*, wrote with little sympathy of a mortally wounded enemy officer who was believed to be the nephew of Cobourg: 'his look, when he was captured, was that of a fierce German tyrant, not daring to look republicans in the eye.'[27] However, a report from the same paper describing a dinner held by the representative on mission, Saliceti, for the English and Spanish officers captured at Toulon provided a sympathetic depiction of the English General O'Hara, not least because of his admiration for French liberty.

> The English officer often repeated with the composure of a man who was deeply moved: 'It is too bad to fight with such people! For our sans-culottes were like brothers sustained by the principles that are the basis of our constitution.'[28]

The report implied, even by its suggestion that it was appropriate for a representative of the French nation to dine and engage in civilized discourse with his captive, that not all enemy soldiers were beyond the pale of civilization.

The report concerning General O'Hara not only gave the enemy a human face, but it also suggested that he should be treated in a humane way. The *Anti-fédéraliste* insisted that 'our soldiers have manifested the sublime heroism of the true Jacobin; they became the nurses and the brothers of the wounded .... The English will see, contrary to what our traitors have led them to believe, that they are not dealing with a rabble of undisciplined brigands.' Nothing could be more disconcerting for the enemy, concluded the journal, than the humanity with which they were treated in spite of their crimes: 'Our brothers are massacred in Toulon; the representatives of the sovereign people are immolated before their eyes, and we see in them only men when they fall into our power.'[29] Another organ of the official Jacobin press, the *Journal de la Montagne*, also insisted upon the humanity of French soldiers toward their enemies. The paper cited an anecdote in which a wounded French soldier had insisted upon a surgeon treating an Austrian prisoner first. To the surgeon's protests that 'He is an Austrian', the Frenchman replied, 'What difference does that make? He is a man.'[30] Such affirmations of generosity toward a human enemy contrasted with the uncompromising message to show no mercy expressed elsewhere in Jacobin propaganda.

---

26 Mittié, *Descente en Angleterre: Prophétie en Deux Actes et en Prose* (Paris, 1798).

27 *Anti-fédéraliste*, 30 vendémiaire, Year Two (21 October, 1793).

28 Ibid., 21 Frimaire, Year Two [11 December, 1793].

29 Ibid.

30 *Journal de la Montagne*, 9 July, 1793.

The Enemy Within

A second factor which served to mitigate the violence of that message derived from the Jacobin preoccupation with conspiracy from within, which tended to elevate internal traitors above foreign enemies in the hierarchy of crime. A speech by a Jacobin of Valenciennes, made on the eve of the declaration of war in April 1792, denied that the Revolution's enemies were to be found in the ranks of the enemy armies.

> All those troops are slaves, subject to the just or unjust intentions of their despotic rulers: they resemble our own who, four years ago, were nothing more than automatons or veritable machines to be moved at will.[31]

The real enemy, he said, were traitors to the fatherland, the foreign despots and all enemies of the constitution. The *Ami Jacques, Argus du Département du Nord*, which reported the speech, a month later included an exhortation to the Army of the North to be vigilant toward 'the traitors, our internal enemies .... Take notice that if we are able to vanquish our internal enemies, those from without will in the same instant be vanquished.'[32]

The rather paradoxical effect of this preoccupation with the enemy within is that it undermined the efforts to demonize the foreign enemy, whose image was strangely enhanced when set against the turpitude of treasonous French generals or the cruelty of French *émigré* troops. During the first year of the war, the worst atrocities were attributed to *émigré* forces. A report in the *Courrier de Strasbourg* on 29 August 1792 was taken up by the *Ami Jacques, Argus du Département du Nord* on 29 August and then widely disseminated in the Parisian press. It recounted the atrocities committed by *émigré* soldiers at a village between Philippeville and Maubeuge.

> They cut the noses and ears off men and the breasts off women, but only after slaking their brutal lusts with them. Not even children were spared. While these monsters occupied themselves with these horrors, they were surprised by a corps of volunteers and troops of the line army who, informed of these abominations, surrounded the village and set fire to it. They did not let a single one of these scoundrels escape, and responded to those who asked for mercy by pushing them back into the fire and saying, 'No mercy in this world, go seek it in the other.' Under the ashes were found twenty-five bodies without heads, which causes one to presume they were refractory priests.[33]

Even if one disregards the last sentence, with its mocking reference to the French clergy who had refused to swear their loyalty to the constitution, this anecdote strains the reader's credulity. As a political parable, however, insisting upon the irreconcilability of revolutionaries and *émigrés* as well as upon the necessity of showing no mercy to the latter, it made its point.

---

31 *Ami Jacques, Argus du Département du Nord*, 17 April, 1792.
32 Ibid., 22 May, 1792.
33 *Ami Jacques, Argus du Département du Nord*, 29 August, 1792.

Republican propaganda often juxtaposed honourable foreign soldiers with dishonourable *émigrés*. Saulnier's play celebrating the defence of Thionville in 1792 represents the Austrian general, Waldeck, as an honourable soldier who spurns the demand by the *émigré* villain, Dautichamp, that he kill the captive son of Thionville's commandant, General Wimpfen. Such demands, says Waldeck, 'would give pleasure to cannibals;/But they must cause horror/To the warrior led by honour.'[34] By insisting upon the distinction between French traitors and enemy soldiers, revolutionary propagandists admitted the possibility that the latter were honourable foes.

The Concept of Honour

This reference to honour points to a third factor that qualified the ferocity of revolutionary war propaganda – the persistence of a conservative ideal of military honour. Of course, at their most extreme, revolutionary publicists repudiated this ideal as an ethic that belonged to the old regime. To Hébert, 'honour' and 'glory' were the unmerited attributes of aristocratic generals, who acquired them through the unheralded sacrifice of their soldiers' lives.[35] Counter-revolutionary propagandists represented the French armies as a rabble devoid of discipline and honour, qualities which had emigrated along with their officers. The royalist Montjoye wrote that 'the true military spirit faded away little by little' as soldiers became infected by revolutionary ideas in the Jacobin clubs.[36] The right-wing *Gazette de Paris*, describing an oath sworn by *émigré* cavalry on the tomb of Marshal Turenne, presented the event as a glowing expression of 'French honour'.[37] Nevertheless, despite its devaluation by the Jacobins and its appropriation by the royalists, the concept of honour as a motivational force never entirely disappeared from the military culture of the Revolution and it underwent a significant renaissance after Thermidor.[38] A pamphlet entitled *The Perfect Warrior, or the Military Spirit*, published in 1792, associated patriotism and honour: 'After patriotism, that source of the ancient virtues, honour is the sentiment that most elevates human nature.'[39] Military honour implied many things: courage, devotion to the state, love of glory and humanity were all connected to the eighteenth century warrior's code. Perhaps the best definition of that code was provided by Andrew Ramsay's life of Turenne, published in 1735, which held up this great seventeenth century marshal as the incarnation of the honourable soldier.[40] Above all, military honour was associated with the idea of professionalism and it implied respect for other professionals, regardless of their national or political

---

34 N.-F.-Guillaume Saulnier, *Le Siège de Thionville* (Paris, n.d.), p. 25.

35 *Père Duchesne*, no. 321.

36 *Ami du Roi*, 6 May, 1792.

37 *Gazette de Paris*, 10 May, 1792.

38 John A. Lynn, 'Toward an Army of Honor: The Moral Evolution of the French Army, 1789-1815', in *French Historical Studies*, 16, 1989, pp. 152-173.

39 Anon., *Le Parfait Guerrier, ou l'Esprit Militaire* (Paris, 1792), p. 25.

40 Andrew Ramsay, *Vie de Turenne*, 2 Vols. (Paris, 1735).

affiliations. Ramsay concluded his study of Turenne by describing the tears shed by Montecuccoli, his opponent in many campaigns, upon learning of his rival's demise.

There is much evidence that this warrior's code was in the ascendant after Thermidor. It was manifest in the Directory's tributes to fallen generals – Duphot, Marceau, Hoche and Joubert – and it culminated in the cult of Latour d'Auvergne, the 'first grenadier of the army', who was killed at Neuberg on 27 June 1800 after many years' service in the armies of both the old regime and the Revolution. The tributes to these soldiers were increasingly apolitical celebrations of military professionalism. Most importantly, the honours accorded the fallen warriors of the Directory united both friend and foe in common gestures of respect. French newspapers described the tears shed by the Austrian General Kray upon the death of General Marceau in 1796. Not only did the Austrians seek to ease their captive's last moments, but they provided a ceremonial guard of honour to escort his remains and his riderless horse as they were returned to the French camp. As the General's body was buried, the Austrian garrison at Ehrenbreitstein fired an artillery salute.[41] Similarly, it is significant that among the tributes promised to Latour-d'Auvergne in 1800 was the erection of a monument on the site where he fell 'consecrated to the virtues and to courage, and placed in the safe-keeping of the brave men of all countries.'[42] The emphasis upon the idea that 'brave men of all countries' could unite to honour the memory of one of their number is evidence for the emergence of a strong sense of professional corporate identity that transcended national boundaries. A sense of that identity is also provided by an encomium to the soldiers of Bonaparte's Army of Italy in the *Courrier de l'Armée d'Italie*:

> Terrible in combat, they are humane and generous in victory. The dying or wounded Austrian finds in the French soldier a protector, a friend, who ministers to him. The same warrior, who has fought and vanquished his adversary, now lifts him up, staunches his wound, sees in him only a man, a brother, a victim of the fortunes of battle …. Thus the young vanquisher of Wurmser [Napoleon Bonaparte, who defeated General Wurmser at Castiglione in 1796] did not hesitate to give homage to the talents and vigour of an old man whose defeat he modestly attributed to the hazards of fortune.[43]

Such descriptions, emphasizing the humane and respectful treatment of fellow professionals, are a far cry from the Jacobin view of the war as a merciless struggle against a bestial and barbarous foe.

The revolutionary discourse on the war was comprised of various competing elements. Three of those elements – on wars of liberation, on the enemy within and on the warrior's professional code of honour – balanced the Jacobin discourse of

---

41 Anon., 'Le Dernier Combat de Marceau et les Honneurs Funèbres Rendus à Marceau', in *Carnets de la Sabretache*, 1899, pp. 272-287.

42 Anon., *Vie Politique et Militaire de Latour-d'Auvergne, Premier Grenadier des Armées Françaises...* (Paris, Year Eight [1799-1800]), p. 15.

43 *Courrier de l'Armée d'Italie*, 15 October, 1797.

hate. The messages conveyed through a diverse array of media to the people and soldiers of revolutionary France were mixed; hatred had its place, particularly during the Terror, but so too did humanity and honour.

Soldiers Bear Witness

Assessing the degree to which soldiers in the revolutionary armies responded to these competing messages is problematic. On an impressionistic or anecdotal level, the testimony of soldiers in their letters and memoirs echoes both the zealotry of Jacobin propaganda and the more detached professionalism of the warrior's code. The *carnet de route* of Sergeant Fricasse paid tribute to the patriotism of his comrades as well as to their antipathy toward the *émigrés*. Citing a verbal exchange between sentries from the opposing sides at the siege of Maubeuge in 1793, Fricasse recorded the words of a volunteer who repudiated his *émigré* brother, whose voice he had recognized: 'You are not worthy to live; you are not a human being, but a true barbarian.'[44] In describing a subsequent engagement with *émigrés* Fricasse stated: 'We took very few of them prisoners, because they did not give themselves up willingly.'[45] More enthusiastically bloodthirsty were the letters of Alexandre Brault, who also expressed particular animosity toward the *émigrés*:

> There are absolutely no cruelties to which they do not resort in their desperation; only death can release them from an unhappy existence for which no-one has sympathy; that is why they fight to the death; I would take great satisfaction in capturing one alive to make him dance at my pleasure.[46]

In a later letter, Brault declared that he had witnessed the summary execution of over 400 *émigrés*.[47] Bastien Belot, writing to his brother, claimed, 'We killed 500 *émigrés* at Rimpsall and burned a company in a mill.'[48] Despite the doubtful authenticity of this claim, the testimony of a surgeon attached to the Army of the Rhine bears witness to the *émigrés* determination to fight to the death, which stemmed from their awareness of the Convention's decree of 29 March 1793 prescribing the death penalty for any Frenchman captured bearing arms against the Republic. Renoult described his efforts to treat 50 *émigré* prisoners who were 'horribly cut about, not having wanted to surrender' and who at first refused treatment in the expectation that they would immediately be shot.[49] Renoult's witness would seem to suggest

---

44 Jacques Fricasse, *Journal de Marche du Sergeant Fricasse de la 127e Demi-Brigade, 1792-1802*, ed. by Lorédan Larchey (Paris, 1882), pp. 16-17.

45 Ibid., p. 23.

46 Ernest Picard, *Au Service de la Nation: Lettres de Volontaires (1792-1798) Recueillies et Publiées par le Colonel Ernest Picard* (Paris, 1914), p. 155.

47 Ibid., p. 171.

48 Ibid., p. 19.

49 Adrien-Jacques Renoult, *Souvenirs du Docteur Adrien-Jacques Renoult* (Paris, 1862), pp. 11-12.

that the resentment expressed toward the *émigrés* by republican troops was more than just patriotic rhetoric rehearsed for home consumption. Insofar as the *émigrés* were concerned, Jacobin antipathy toward these 'cowards who have abandoned their homes to revolt against their fatherland'[50] does indeed appear to have intensified the bitterness of military conflict.

On the other hand, there are also examples of soldiers who were deaf to the revolutionary message – soldiers like Pion de Loches, who avoided conscription three times before finally the *levée en masse* forced him into service. Gradually, he acquired a taste for the military life – 'I began to adopt military attitudes and saw the possibility to make my way in the profession of arms.'[51] To Pion de Loches, the army was a profession rather than a patriotic calling.

Nor were all relations with the enemy marked by ideological tension. At a very basic level, tacit truces were sometimes observed in order to allow both sides to satisfy their needs without impediment. *Canonnier* Bricard described how, following the Battle of Altenkirchen (4 June 1796), French and Austrian forces faced each other across the Lahn: 'It was forbidden, on both sides, to fire shots, so that men and horses might make use of the water.'[52] Pierre Girardon described a similar situation along the Rhine in early 1794.[53] More formal truces could lead to substantial fraternization. Alexandre Boutrouë, a brigadier serving in the Army of the Rhine and Moselle in Germany, wrote to his brother in October 1797 that he had recently dined with Austrian officers in Manheim, but that at the very moment he was writing, an order had been received 'that all communication with the enemy must cease' and consequently 'no Austrian officer can any longer be received in our cantonments.'[54] Boutrouë's letter bears witness to a significant level of interaction between French officers and their Austrian counterparts. There is little sense of the ideological gulf between them imagined by Robespierre.

Armies of Occupation

Even more difficult to assess than the attitude of troops toward enemy soldiers is the question of how their treatment of civilians was conditioned. In principle, of course, humane treatment of foreign civilians was prescribed both by Jacobin ideology and by the military code of honour. Instances when revolutionary forces had clearly violated the rights of civilians, such as Jarry's action at Courtrai, were deplored in the revolutionary press. Gorsas, in the *Courrier des Départemens*, was scandalized by Jarry's 'execrable action': 'The soldiers of liberty must punish the traitors; but

---

50 Picard, *Au Service*, p. 171.

51 Pion de Loches, *Mes Campagnes (1792-1815)* (Paris, 1889), pp. 22-23.

52 Bricard, *Journal du Canonnier Bricard*, ed. by Lorédan Larchey (Paris, 1891), p. 191.

53 Pierre Girardon, *Lettres de Pierre Girardon, Officier barsuraubois pendant les guerres de la Révolution (1791-1799)* (Bar-sur-Aube, 1898), p. 41.

54 Alexandre Boutrouë, *Lettres d'un chef de brigade*, ed. by M.A. d'Hauterive (Paris, 1891), p. 32.

to go torch in hand to burn the properties of citizens and thereby lump together the guilty and the innocent is an indignity, an execration.'[55] Furthermore, harsh penalties were prescribed for crimes committed by soldiers against civilians. A law of 27 July 1793 specified the death penalty for rape or pillage. The revised penal code of 1796 also specified the death sentence for pillage and a penalty of eight years in irons for rape. During the Terror, at least, the military courts were severe in their application of the law, ordering the execution of soldiers who went aside to plunder.[56]

Nevertheless, even during the early days of revolutionary enthusiasm, the conduct of soldiers in occupied territory, by their own admission, was rarely beyond reproach. One, Demonchy, referred with a certain unconscious irony to the 'humanity' with which he and his comrades treated their hosts in occupied Belgium in 1794:

> We took horses, chickens, cows, sheep, pigs, that is to say all livestock in general, and, after these expeditions, we went to the peasant's home and, after feeding well and drinking his wine, we f ... his women and carried off his crowns and shillings. That is how we behave in the homes of the Imperial Gentlemen. They do much worse in our homes; they behave with cruelty, while we Frenchmen always act with humanity.[57]

Much worse was to come, in 1795 and 1796, as the revolutionary armies advanced into Germany and Italy. The rapacity of Generals Jourdan's and Bonaparte's armies was so great that popular uprisings occurred which endangered military operations. In Italy, Bonaparte sought to control the situation through exemplary violence, executing nine soldiers to deter pillage and unleashing a punitive attack on the city of Pavia, which was subject to rape and pillage for 24 hours. In Germany, Jourdan and his generals vainly sought to preserve order and discipline. The military justice files at Vincennes for the Army of the Sambre and Meuse are overflowing with cases of desertion, assault, drunkenness, theft, rape, murder and pillage from the campaigns of 1795 and 1796.[58] The correspondence of General Collaud provides a vivid picture of the situation. Recounting his actions as the summer campaign of 1796 began, Collaud promised that 'terrible examples will be made during the first days on the march.'[59] He was as good as his word, executing a soldier for striking an officer and marching the division past the victim's body, 'to serve as an example'.[60] Such measures, as Collaud himself admitted, were in vain. On 24 July, he wrote in

---

55 *Courrier des Départemens*, 4 July, 1792.

56 See, for example, the records of the military tribunal set up in Strasbourg by St. Just and Lebas to impose discipline on the Army of the Rhine, in Antoine-Vincent Legrand, *La Justice militarie et la Discipline à l'Armée du Rhin et à ll'Armée de Rhin-et-Moselle (1792-1796)* (Paris, 1909), pp. 80-95; Georges Michon, *La Justice Militaire Sous la Révolution* (Paris, 1922); Georges Michon, 'La Justice Militaire sous la Convention à l'Armée des Pyrénées Orientales,' in *Annales Historiques de la Révolution Française*, Vol. 3 (1926), pp. 37-46; Peter Wetzler, *War and Subsistence: the Sambre and Meuse Army in 1794* (New York, 1985), p. 244.

57 Picard, *Au Service*, 142.

58 Archives de Guerre, B1-311 to B1-316.

59 Archives de Guerre, B1*232* letter of 11 prairial, Year Four (30 May, 1796).

60 Archives de Guerre, B1*232* letter of 11 messidor, Year Four (29 June, 1796).

despair that if the depredations of his division continued he would ask to be relieved of his command: 'I have absolutely no desire to dishonour myself at the head of such brigands, who burn, steal, murder, rape, commit every sort of crime and do not even deserve to be called men.'[61] On marches, he claimed, as many as 5,000 men would leave the column 'to commit all sorts of crime' in the villages along the route, resisting with force of arms officers who sought to bring them to order. Instead of arriving in camp with a division, he sometimes arrived with little more than a brigade.

Revolutionary publicists continued to insist upon the honourable conduct of the armies. The *Courrier de l'Armée d'Italie* declared that 'the republican heroes have not sullied, through odious vices, dishonourable crimes, or through criminal abuse of their warlike virtues and their stunning successes the noble qualities they seem to have acquired from their predecessors and their models.'[62] 'Terrible and salutary examples' had been made of the few miscreants, claimed the journal, who were precipitated like 'modern Manliuses' from the Tarpeian rocks. Such exalted prose bore little relation to the cruel and mundane reality of war in occupied Europe.

Various explanations can be offered for the depredations of the revolutionary armies under the Directory. The generals themselves pointed to the obvious failure of the commissariat to provide adequate supplies, which effectively forced soldiers to resort to theft in order to survive. General Collaud's letter of 6 Thermidor concluded by stating that the want of supplies was the 'source of all the disorders'. The testimony of soldiers themselves confirmed that their actions were determined by base necessity. Jean Chatton, a soldier in the Army of the Sambre and Meuse, described the famine of the army besieging Coblentz in 1794-5. To survive, he was forced to walk five leagues in the search for potatoes or apples, and to boil clover in his *marmite*. Despite narrowly escaping execution at Coblentz for marauding, when his unit was transferred to the blockade of Mainz, 'we thoroughly pillaged the poor peasants in order to get something to eat.'[63]

Without denying such material imperatives, some historians have suggested that the increasingly militarist culture of the army contributed to the misconduct of its troops under the Directory. Georges Michon deplored the militarization of justice within the army, which became a tool to enforce the authority of officers over their men rather than to protect civil society.[64] From this perspective, the increasing professionalization of the army[65] and its growing self-awareness as an institution set apart from and in some respects superior to civil society, might be presumed to have diminished soldiers' inhibitions concerning the exploitation and abuse of civilians. It

---

61 Archives de Guerre, B1*232* letter of 6 Thermidor, Year Four (24 July, 1796).

62 The *Courrier de l'Armée d'Italie*, 23 October, 1797.

63 M.E. Gridel and Capitain Richard, eds, *Cahiers de Vieux Soldats de la Révolution et de l'Empire* (Paris, 1903), pp. 9-11.

64 Michon, *Justice Militaire*.

65 On this theme, see Rafe Blaufarb, *The French Army, 1750-1820: Careers, Talent, Merit* (Manchester, 2002).

is also possible that the increasing oppression of military discipline within the army may have led to the displacement of violence as soldiers brutalized by their officers in turn brutalized civilians.

The Army of the Sambre and Meuse

The military justice records for the Army of the Sambre and Meuse lend some support to this argument. The records of Years Four (1795-96) and Five (1796-97) contrast with those of Year Two (1793-1794). Under the National Convention, with the representatives on mission to the army playing a leading role, crime against civilians (rape or pillage) was punished severely while military crime (abandoning one's post, insubordination) was viewed with relative indulgence. Under the Directory, this balance was reversed. In August, 1794, Captain Gabriel Lijard, convicted of encouraging his soldiers to pillage, was sentenced to be shot in front of his battalion and the sentence was to be read to each company 'in order to inspire more and more fear in those who might be tempted to follow his example.'[66] In June, 1796, a Sergeant Maffré, convicted of negligence in failing to prevent pillaging, 'which he had seemed to authorize by his presence' in the rear of the army, was merely stripped of his rank and given one month in prison.[67] In December of that year, Adjutant-General Bonami, although convicted of exploiting civilians on a grand scale, imposing arbitrary taxes in money and goods and taking hostages in Hamburg and three other towns, was sentenced only to two years in prison.[68] These examples illustrate the growing tolerance of military courts for crimes against civilians. Other cases demonstrate the declining tolerance for military crime. On 13 July 1794, a Citizen Deprat, responding to a blow from his officer with the threat 'that if he did it again he would hit him between the legs with his soup-pot' was acquitted and returned to service.[69] On 26 December 1796, François Bray was not so fortunate. Accused of assaulting his officer, Bray was convicted and sentenced to death despite his evidence that he was so drunk at the time that he had no recollection of the event and despite the suggestion at least that the officer – also drunk – had struck the first blows.[70]

Yet too much should not be made of this evidence. Despite its change of emphasis, military justice in the Army of the Sambre and Meuse under the Directory did not endorse a regime of rape and pillage. Exemplary executions of soldiers convicted of pillage continued to take place, as four soldiers of the 96th and 105th demi-brigades discovered to their cost when they were sentenced to death on 3 July 1797. Furthermore, when actual violence was done to civilians the courts took a serious view of the matter, imposing the death penalty for murder or attempted murder.

---

66 Archives de Guerre, B1-311.
67 Archives de Guerre, B1-312, 25 prairial, Year Four (13 June, 1796).
68 Archives de Guerre, J2-335.
69 Archives de Guerre, B1-311.
70 Archives de Guerre, J2-335.

Thus, Henry Charbonnier was sentenced to death on 16 July 1797 for a brutal sword attack on two women and a domestic in an auberge at Dierdorff.[71] Jean-Baptiste Gernemont was also sentenced to death for terrorizing a widow and her family in a mill near Linsik. Gernemont had assaulted the woman, her two daughters and their domestic, wounding the girls with his bayonet. He also shot the family's dog.[72] Such extremes of violence were severely punished by the courts, which were otherwise inclined to leniency, often taking into account exceptional circumstances: notably the hardship which drove soldiers to steal as well as drunkenness.

The generals commanding the Army of the Sambre and Meuse themselves were well aware that the good order of the army depended upon orderly relations with the civilian population it encountered. General Jourdan warned his troops 'that success depends upon the order and subordination that reign within the army; that the pillage to which they have often abandoned themselves has the most harmful consequences and that it prevents the army from being able to subsist.'[73] He therefore exhorted his generals, after promising to destroy the villages of peasants who fired on their men, 'at the same time to insist upon the greatest discipline from the troops and to prevent all pillage.'[74] To General Collaud, he sent the warning, 'It has long been my experience that it is by sending troops on requisitions that discipline is lost and pillage introduced.' He insisted that requisitions should be led by officers who would be responsible for the conduct of their men, 'and that you will punish or have judged by military courts those who do not do their duty.'[75]

The correspondence of General Kléber is particularly interesting for what it reveals about the values of the army's high command. Referring to the depredations of General Bonami, Kléber's instruction to General Ligneville was: 'Strike, my dear General, all individuals who by their conduct dishonour the French name and use all the means in your power to stop the exactions.'[76] The reference to honour was even more explicit in a proclamation addressed by Kléber jointly to the soldiers under his command and to the local inhabitants. Kléber condemned the 'armed scoundrels wearing French uniform' who had imposed exactions on the peasants and provoked resistance that had necessitated the burning of the village of Schwabenheim. The honour of the army and of the fatherland, he said, required him to seek out and punish those responsible, 'in order to purge the army of the brigands who dishonour it.' Finally, he defined the standard of conduct to be expected of his army:

> It is necessary that the warriors who compose it, far from being an object of fear to the innocent and to the peasant, should instead be their protectors; that, worthy of the exploits by which the Army of the Sambre and Meuse has so often come to Europe's attention,

---

71 Archives de Guerre, J2-335.

72 Archives de Guerre, J2-335, 25 nivôse, Year Five (14 January, 1797).

73 Archives de Guerre, B1*224*

74 Archives de Guerre, B1*224*, letter to General Ernouf, 30 brumaire, Year Four (21 November, 1795).

75 Archives de Guerre, B1*232*, 19 nivôse, Year Four (9 January, 1796).

76 Archives de Guerre, B1*219*, 18 brumaire, Year Five (8 November, 1796).

we prepare ourselves to march with confidence against the enemy, to fight him and to vanquish him.[77]

The proclamation directly appealed to the army's dignity and pride, insisting that the criminal actions of the 'scoundrels' and 'brigands' were an affront to its honour. A proclamation to General Championnet's division, offering 3,000 new uniforms to its officers, made a similar contrast between the behaviour of brigands and the behaviour of soldiers.

> The General-in-Chief has seen in the ranks officers without shoes and soldiers without clothing. The dilapidated appearance of our battalions has caused his heart to bleed. But their proud and martial bearing has revived his courage. Their rags and their uniform he saw as titles of honour, certificates of probity, and he will always keep his heart's eye upon those brave men who have resisted the temptations of luxury and brigandage in order to remain in their ranks to defend their flags. These faithful friends of honour are also his and never will his paternal care lose them from sight.[78]

Making a virtue of necessity, the proclamation reminded the officers of what it was that distinguished them, as soldiers, from brigands. It was their 'proud and martial bearing', their 'probity', their devotion to duty and above all their love of honour that set them apart.

This was an important reminder at a time when the Army of the Sambre and Meuse, by all accounts, had largely lost its martial appearance along with its discipline. A contemporary description of the French armies in Germany stated that 'their equipment was truly bizarre' and that soldiers were dressed in bed sheets, carpets, multi-coloured coats, women's underwear as well as choir gowns and surplices stolen from churches: 'It was, so to speak, a mascarde.'[79] Contemporary German prints also emphasized the unmilitary dress and comportment of French troops. The generals' proclamations, even as they recognized the importance of the uniform as a symbol, insisted that it took more than a uniform to make a soldier – it also took a sense of honour.

That sense of honour may appear to have been a feeble counterweight to the circumstances that determined the progressive degeneration of General Jourdan's army in 1795 and 1796. Beyond the debatable merits of exemplary punishment as a deterrent, however, it is about all that stood in the way of complete dissolution. Furthermore, there is evidence that some soldiers, at least, internalized the values of this warrior's code and that it did indeed condition their conduct. *Canonnier* Bricard's memoir drew a clear distinction between the 'true soldiers' who applauded General Jourdan's efforts to suppress brigandage and the 'scoundrels' who raped and pillaged.

---

77 Archives de Guerre, B1*220, 14 vendémiaire, Year Five (5 October, 1796).
78 Archives de GuerreB1*222, 15 vendémiaire, Year Five (6 September, 1796).
79 *Carnets de la Sabretache* (1902), pp. 135-148.

It will be difficult to believe that some soldiers, unworthy of the name, engulfed the houses to break into barrels of wine, to steal, to rape the women, while the true soldiers were hacked about as they protected the retreat. Will it be believed that many were taken by the enemy, some in wine up to their knees, the others raping mothers in the presence of their husbands![80]

Bricard insisted that there were few pillagers in his unit. He confiscated pillaged goods and restored them, when possible, to their rightful owners.[81]

The Warrior's Code of Honour

The memoir of Bricard, echoing the proclamations of his generals, suggests that, in the new age of democratic warfare, the warrior's code of honour remained an important, though fragile, bulwark against the barbarization of warfare. The French Revolution's mass mobilization of society and representation of war as a conflict of irreconcilable ideologies threatened to sweep aside enlightenment constraints on warfare. Atrocity stories representing a bestial and barbarous foe were calculated to justify a 'no-holds-barred' warfare, in which traditional scruples about the use of heated shot or the slaughtering of defeated enemies were ignored. The warrior's code of honour survived as a significant counterweight to this propaganda, which was itself undermined by the contradictions of Jacobin ideology. 'True soldiers', as Bricard insisted, did not rape and pillage and they treated a defeated enemy with respect and humanity.

The experience of the French Revolution would seem, therefore, to support the argument applied by Michael Ignatieff to the modern world that the warrior's honour, although 'a slender hope ..., may be all there is to separate war from savagery.'[82] The at first creeping and then rampant militarism of French society as it passed from its revolutionary to its Napoleonic periods is easy to deplore, as are its wars of conquest which ultimately led to the domination of the European continent. Nevertheless, the moderating effect of the professional code of honour which came to prevail in the French armies must also be acknowledged. In our own times, when irregular warfare and extreme ideological conflict repeatedly burst asunder the laws of war, the channelling and restraint of violence by regular armies animated by the warrior's code of honour provides one means of limiting barbarity in warfare.

---

80 Bricard, *Journal*, p. 237.
81 Ibid., p. 209.
82 Michael Ignatieff, *The Warrior's Honour: Ethnic War and the Modern Conscience* (Harmondsworth, 1998), p. 157.

Chapter 4

The British Army and Controlling Barbarization during the Kenya Emergency

Huw C. Bennett

Omer Bartov introduced the term 'barbarization' in *The Eastern Front, 1941-45, German Troops and the Barbarisation of Warfare*, arguing that it resulted from conditions at the front, the social and educational background of the junior officers, and the political indoctrination of the troops.[1] Without producing an explicit definition, Bartov's analysis views it as a process with multiple causes, best seen in action from below. This chapter agrees with Bartov in seeing it as a process with many causes but slightly adapts the view from below, to see how ordinary soldiers interacted with orders from above. Also, Bartov includes environmental factors in his analysis, such as the terrain and the prevailing ideology. The focus here is instead upon factors substantially controllable by the Army as an institution, looking at the organization's ability to control its own soldiers. While there are indeed many causal influences in the highly complex barbarization process, this chapter focuses upon discipline. Because the British Army fighting a counter-insurgency in Kenya between 1952-56 possessed no genocidal aims or strategy of war crimes, the focus must instead be upon how the disciplinary system sought to stop atrocities happening. The functionalist approach adopted here is interactional, allowing an analysis which asks not only what the commander ordered from top to bottom, but equally how ordinary fighting men conducting operations imposed their conception of the conflict up towards the commander.[2] The commander's formal intentions were not

*   The author expresses his thanks for assistance and encouragement from Professors Colin McInnes and Martin Alexander at Aberystwyth, Professor John Ferris at Calgary, the UK Economic and Social Research Council, and the staff at the National Archives, Kew and the Imperial War Museum, London. References made to the Papers of General Sir George Erskine with kind permission of the copyright holder and the Trustees of the Imperial War Museum.

1   Omer Bartov, *The Eastern Front, 1941-45, German Troops and the Barbarisation of Warfare*, (Basingstoke, 1985).
2   Tony Ashworth, *Trench Warfare, 1914-1918: The Live and Let Live System*, (London, 1980); Leonard V. Smith, *Between Mutiny and Obedience: The Case of the French Fifth Infantry Division during World War I*, (Princeton, 1994).

always identical with the way force was applied on the ground, an outcome due not only to chance and the enemy's activities, but also to a generally under-emphasized dimension – the Army's own social structure. In examining the limitations to official strategy the roles played by the Army's component parts helps explain the prominent position taken throughout the campaign by informal actions which occurred without permission, or even against policy. Atrocities represented not only an avoidable breakdown in discipline within certain units, but also paradoxically often logically stemmed from flaws in policy. Constraints upon the command's power over its own forces meant that the ideal vision of British counter-insurgency as restrained and conducted according to the minimum force ethical principle were compromised. An ongoing disciplinary struggle within the Army reflected the entire war's position as a Kenyan civil war. The Emergency as a whole was a complex civil war within the Kikuyu tribe and against settler and loyalist African factions.[3] For the Army, conflict against the Mau Mau insurgents coincided with dissent between British battalions, King's African Rifles (KAR) battalions, the Kenya Regiment, and also the non-military state security forces such as the Police, Home Guard and the Provincial Administration.

This horizontal interaction coexisted with the vertical, hierarchical command structure's constant compromises with strategic ideals and social realities. The Army simultaneously waged a war in which restraint and excess were evident. This chapter considers how far the Army command actually controlled the behaviour of the troops and in what ways the troops enforced their own vision of the war onto the command. On the one hand, General Erskine exercised a tight grip over his forces in demanding they treat the populace with restraint. Conversely though a barbarized form of war fighting in Kenya had rapidly developed over the opening nine months. As a result of it the Army command never fully achieved its aim to fight a restrained war and had to concede to its own men's desires for violence. Let us begin by examining the ways in which the Army sought to prevent barbarized warfare by protecting non-combatants.

---

3   David Anderson, *Histories of the Hanged. Britain's Dirty War in Kenya and the End of Empire*, (London, 2005), pp. 9-54 on the factions in the civil war. Maloba emphasizes the anti-colonial dimension, whereas Furedi sees Mau Mau as essentially an agrarian revolt. For Berman the revolt was intrinsically a class struggle. Despite disagreement over Mau Mau's origins, which is not a major concern here, there is general agreement that contemporary European concerns about a reversion to barbarism were wrong – a theme eloquently expounded by Lonsdale. Wunyabari Maloba, *Mau Mau and Kenya: An Analysis of a Peasant Revolt*, (Bloomington, 1993), p. 1; Frank Furedi, *The Mau Mau War in Perspective*, (London, 1989), p. 7; Bruce Berman, *Control and Crisis in Colonial Kenya. The Dialectic of Domination*, (London, 1990); John Lonsdale, 'Mau Maus of the Mind: Making Mau Mau and Remaking Kenya', *Journal of African History*, 31, 1990, pp. 393-421.

The Disciplinary Situation: 'Restraint Backed by Good Discipline'

During the Emergency the security forces were subject to either military or civil law, and in the Army's case, to both.[4] Operating within legal parameters meant that officers from the Commander-in-Chief on down had to closely watch the discipline in their units. This was vital given the difficulty with distinguishing between combatants and civilians, and the need to control men fighting against an enemy notorious for provocative, brutal massacres. Throughout the conflict disciplinary measures were taken to prevent atrocities by the Army, although the period between 23 June 1953 and 11 March 1954 marked the most intensive period. On 17 April 1953, Governor Baring condemned 'acts of indiscipline involving the unlawful causing of death or injury, the rough handling of members of the public, suspects, or prisoners.' His communiqué emphasized how the security forces had already been issued with instructions on restraint, that all complaints were investigated and where sufficient evidence existed, prosecutions enacted; though these were naturally insignificant in number.[5] Therefore, the disciplinary measures taken in the nine month period from June 1953 were not without precedent, for whatever criticisms arose over the security forces' conduct in the opening eight months, they were not engaged in all-out killing sprees against the Kikuyu population.

General Sir George Erskine took command of the newly-formed, independent East Africa Command on 7 June 1953 after a catastrophic single night left 74 slaughtered in Lari village, and prisoners and arms humiliatingly liberated from Naivasha police station.[6] Whether from personal conviction or the logical recognition that stricter discipline would pay strategic dividends, Erskine soon decided to impose his mark on the forces.[7] On 23 June all officers received a message:

> ... the Security Forces under my command are disciplined forces who know how to behave in circumstances which are most distasteful.
> ... I will not tolerate breaches of discipline leading to unfair treatment of anybody.
> ... I most strongly disapprove of 'beating up' the inhabitants of this country just because they are the inhabitants. ... Any indiscipline of this kind would do great damage to the reputation of the Security Forces and make our task of settling MAU MAU [sic] much more difficult. I therefore order that every officer in the Police and the Army should stamp on at once any conduct which he would be ashamed to see used against his own people.

---

4   National Archives, Kew: WO 32/15556: Personal and confidential letter, Erskine to Chief of the Imperial General Staff (CIGS), 9 July 1953.

5   WO 32/21721: McLean Court of Inquiry Exhibit 29: Press Office Handout, dated 17 April 1953.

6   Anderson, *Histories of the Hanged*, p. 128, p. 178.

7   Anderson opts for the former, whilst Percox views improvements in the command structure, and thus conceivably also the disciplinary structure, as inevitable once Whitehall decided to install a new commander, regardless of who the individual was. There is no reason why these views should be mutually exclusive. Anderson, *Histories of the Hanged*, p. 261; David A. Percox, 'British Counter-Insurgency in Kenya, 1952-56: Extension of Internal Security Policy or Prelude to Decolonisation?' *Small Wars and Insurgencies* 9/3, 1998, pp. 75-76.

... Any complaints against the Police or Army which come from outside sources will be referred to me immediately on receipt and will be investigated.[8]

Further measures were taken to ensure everyone received and understood the new commander's views. All arriving troops were issued with the order.[9] Along with re-structuring the command and bringing in combat reinforcements, the internal affairs investigative body, the Special Investigations Branch (SIB) was strengthened during the second half of 1953.[10] In general, the 23 June order was obeyed by the Army thanks to tradition, discipline, awareness of the practical benefits of good conduct and the knowledge that Parliament kept a close eye on the situation in Kenya.[11] To further guarantee everyone understood how to maintain a disciplined stance when dealing with the Kikuyu and win the conflict, GHQ issued Operational Intelligence Instruction Number 4 on 1 July. The instruction addressed the correct procedure for dealing with 'Mau Mau prisoners and surrendered personnel', meaning both captured insurgents but more frequently, suspected civilians. Prisoners received an immediate tactical interrogation from whoever captured them to produce actionable intelligence. They were then handed over to the Police as soon as possible, normally within 24 hours, and exceptionally within 72 hours if the informant could lead a patrol to an insurgent hideout. On interrogations, the instruction warned that 'violent methods seldom produce accurate information.'[12] A further order from 70th Infantry Brigade on 24 September directed commanders to ensure all ranks understood that violence was not to be offered to any Kikuyu in the Reserves, including prisoners.[13] The belief that humane treatment would reap better intelligence permeates all these documents.

Another intelligence-related question arose in August when an established practice came to light, one apparently also common in Malaya.[14] It must have surprised some then when Erskine banned the security forces from chopping off the hands from dead bodies in order to fingerprint them and provide information on who had been killed. GHQ's Training Instruction Number 7 stipulated that: 'Under NO circumstances will bodies be mutilated, even for identification.' Bodies were to be removed to the nearest police station where possible, and when the terrain proved impenetrable,

---

8   WO 32/21721: Exhibit 5, 'Message to be distributed to all officers of the Army, Police and the Security Forces', GHQ Nairobi, 23 June 1953.

9   Anthony Clayton, *Counter-Insurgency in Kenya. A Study of Military Operations against Mau Mau*, (Nairobi, 1976), p. 38.

10   A.V. Lovell-Knight, *The Story of the Royal Military Police*, (London, 1977), p. 288.

11   Clayton, *Counter-Insurgency in Kenya*, p. 40.

12   WO 32/21721: Exhibit 4, 'Operational Intelligence Instruction No. 4', dated 1 July 1953.

13   WO 32/21721: Exhibit 12, 'Discipline of Security Forces on Operations', 70th Infantry Brigade order to 3rd, 5th, 7th, 23rd KAR and East Africa Armoured Car Squadron Mobile Column A, 24 September 1953.

14   WO 32/21720: Mau Mau operations: Court's report and findings [hereafter McLean Proceedings], p. 376. A Major Morgan, formerly in Malaya, was reported as imparting the practice at the East Africa Battle School, according to Major R.K. Denniston, 1st Black Watch.

full fingerprints were to be taken from the corpse.[15] This GHQ general order in fact post-dated a number of local directives, such as that put out by 39th Infantry Brigade on 1 July, directing all enemy dead buried and specifically prohibiting the removal of limbs.[16] The issue shows how quickly and thoroughly the Army responded to the new commander's imposition of a new set of military norms, banning a practice commonplace when he arrived and considered a military necessity. It demonstrates the fluidity of 'necessity' and the ease with which alternative, and to Erskine and the British press' sensibilities, less barbarous, practices could be implemented. Although his personal papers do not reveal his thinking on the issue, Erskine possibly realized that mutilation of the dead constituted a war crime under the Hague and Geneva Conventions, and won the Army few plaudits at home, abroad or in Kenya itself.

The testimonies given at the McLean Court of Inquiry in December reveal several new perspective on the chopping off of hands and the replacement fingerprinting policy. Firstly, as mentioned, it was the general practice to chop the hands off an unrecoverable corpse, a 7th KAR officer even describing it as 'a sort of order'.[17] Captain Russell, also serving with the 7th KAR since October 1952, blamed the police, who wanted hands for identification purposes.[18] This explanation is plausible given that by the end of 1953 the Criminal Records Office held fingerprint slips for 475,884 people, so there were records to check against.[19] Another six witnesses concurred on the practice's prevalence in the early days.[20] Brigadier Tweedie, who banned it in his brigade three months after arriving, believed it was done 'not as bestiality but simply because they had no alternative.'[21] With Erskine's intervention though a simple enough alternative appeared. At the Inquiry, 34 witnesses expressed positively knowing the practice was banned.[22] Of these, 24 mentioned carrying the fingerprinting kits as prescribed by GHQ. This evidence suggests the Army command succeeded in imposing discipline even when it went against a procedure established for nine to ten months.

---

15 WO 32/21721: Exhibit 8, 'Training Instruction No. 7, Operations against the Mau Mau', issued by GHQ, 1 August 1953.

16 WO 32/21721: Exhibit 25, '39 Inf Bde Jock Scott Op Instr No. 7', 9 July 1953. The constituent battalions on this date were 1st Buffs, 1st Devons, 1st Lancashire Fusiliers, Kenya Regiment, 4th KAR, and 6th KAR (including a component seconded from 26th KAR).

17 McLean Proceedings, p. 102 (Major J.A. Robertson, 7th KAR).

18 McLean Proceedings, p. 276 (Capt. H.C. Russell, 7th KAR).

19 David Throup, 'Crime, Politics and the Police in Colonial Kenya, 1939-63', in David M. Anderson and David Killingray (eds), *Policing and Decolonisation: Politics, Nationalism and the Police, 1917-65*, (Manchester, 1992), p. 146.

20 McLean Proceedings, p. 223: Lt.-Col. L.W.B. Evans, 5th KAR; p. 227: Major W.E.B. Atkins, 5th KAR; p. 331: Major M.J. Harbage, 4th KAR; p. 393: Brig. J.W. Tweedie, 39th Infantry Brigade; p. 399: Major W.B. Thomas, 39th Infantry Brigade; p. 403: Capt. J.W. Turnbull, Kenya Regiment.

21 McLean Proceedings, p. 393: Brig. J.W. Tweedie, 39 Infantry Brigade.

22 McLean Proceedings. Not all witnesses were asked about it.

However, there were a few exceptions to the rules by criminal elements in the King's African Rifles and the Kenya Regiment.[23] Their actions were not prevented by the orders analyzed above, so the Army sought to see they committed no further infractions. By the same method, the Army wished to clarify to all other soldiers what was unacceptable and provide a deterrent. The method was General Court-Martial, and the showdown took place in late November 1953.

Before the trial began, Erskine wrote to his wife describing how he would shortly try the 'most revolting and unforgivable case', predicting in consequence 'a most violent outcry.'[24] He was right to think the trial would prove sensational. Whilst on a sweep operation in June 1953, Captain Gerald Griffiths pulled up to a 'stop' post manned by two askaris. Finding three prisoners in their custody he asked why they had not killed these forestry workers, before promptly sending one of them on his way. The other two men were handed back their passes, told to proceed and then shot in the back at close range by Griffiths with a Bren machine gun. Returning to the scene some half an hour later to find one man still alive writhing on the road, Griffiths shot him dead at close range with his sidearm after the recently arrived Company Sergeant-Major Llewelyn refused an order to do so. At his trial, the two askaris and a Captain Joy, also present at the time, testified to seeing him shoot the men – who were not running away as the accused claimed. Llewelyn saw him kill the man with a pistol, a charge Griffiths accepted without quibble.[25] But on the ridiculous technicality that the prosecution could not prove this man's identity, Griffiths was acquitted. As one of the Emergency's staunchest contemporary critics pointed out, the prosecution for some unknown reason only pressed a charge for one murder, and through incompetence the murder of another was proven. Arguably, Griffiths should straight away have stood trial for murdering an unknown man.[26]

General Erskine was astonished at the Court-Martial's outcome.[27] Several paths were followed to guard against any possible misreading of the acquittal, arising both from parliamentary pressure on the Secretary of State and Erskine's own desire to see his vision safeguarded. Five days later, the 23 June order was reissued largely unchanged, stressing the commander's determination to 'catch and punish' those who were 'taking the law into their own hands and acting outside my orders.'[28] On 5 December the C-in-C assembled all officers ranking from Lieutenant-Colonel upwards (and their police equivalents) at a special meeting to ram home the issue. In short, the order was deadly serious, would be fully implemented and strictly interpreted by all commanders on the spot, as he could not predict 'every possible stupidity'.

---

23 Clayton, *Counter-Insurgency in Kenya*, p. 40.

24 Imperial War Museum Documents Department, General Sir George Erskine Papers, Accession No. 75/134/1; box 1, file 4: letter dated 27 October 1953.

25 WO 71/1218: Proceedings of the General Court-Martial of G.S.L. Griffiths, for murder, 25 to 27 November 1953.

26 Peter Evans, *Law and Disorder, or Scenes of Life in Kenya*, (London, 1956), p. 262.

27 IWM Erskine Papers, letter to his wife, dated 28 November 1953.

28 WO 32/21721: Exhibit 6: 'Message to be distributed to all members of the Army, Police and the Security Forces', GHQ, 30 November 1953.

The officers were informed about the impending Court of Inquiry and instructed to cooperate.[29] The meeting's immediate impact may be seen in Lieutenant-Colonel Glanville's order to his battalion, commanding his officers to explain Erskine's order to all British and African non-commissioned officers and other ranks. The strongly worded directive probably reflects the gravity adopted by Erskine at the meeting, as Glanville not only passed on the gist but added his own desire to protect the battalion's reputation and '... anyone, be he British or African, who dirty's it will have no mercy from me.'[30]

70th Infantry Brigade, responsible at this point for 3rd, 5th, 7th and 23rd KAR, sent out a comparable order on 8 December. Similarly intending to disseminate the C-in-C's views to all ranks, Brigadier Orr reminded his soldiers how to deal with prisoners, not to engage in killing competitions nor shoot persons out of hand, in slightly more encouraging tones. After all, it was '... a task for which we are all trained, and is not difficult.'[31] One further event happened before the Court of Inquiry which acted to support Erskine's desire to see his forces 'play to M.C.C. rules'.[32] At present there is little information on the General Court Martial of Sergeant Pearson and Private Taylor, two Kenya Regiment soldiers both convicted on 10 December for 'assaulting an African and maliciously burning a house and occasioning bodily harm', with Pearson alone further convicted for 'maliciously wounding an African'. Pearson's award for these activities was a year in prison, while Taylor was sentenced to nine months.[33] This outcome may be interpreted as a victory for the Army's corporate resolve to field disciplined forces subject to the rule of law.

Erskine posted letters to each formation commander requesting a personal assurance on conduct. Affirmative replies arrived on his desk within the week.[34] Also on 30 November, Attlee led Opposition questions in the Commons asking the Government whether they would call an Inquiry into allegations arising from the Griffiths trial, specifically the offering of monetary rewards for kills and competition

---

29 WO 32/21721: Exhibit 7: 'Record of an address made by C-in-C at GHQ East Africa at 1000 hrs Sat 5 December 1953'.

30 WO 32/21721: Exhibit 9: 'Conduct of Security Forces on Ops.', issued by CO 6th KAR, 7 December 1953.

31 WO 32/21721: Exhibit 13: 'Discipline', issued HQ 70th Infantry Brigade, 8 December 1953.

32 The phrase comes from a letter to his wife: IWM Erskine Papers, letter dated 30 September 1953.

33 WO 32/21721: Exhibit 22: 'List of cases brought to the notice of GHQ East Africa in which members of the Military Forces have been charged before Civil Courts, or Courts Martial, or Summarily for offences against Africans', compiled by Assistant Adjutant General, GHQ.

34 WO 32/15834: Letter dated 30 November 1953 from Erskine to Brig. Tweedie (39th Infantry Brigade), Brig. Taylor (49th Infantry Brigade), Brig. Orr (70th Infantry Brigade), Major Huth (Armoured Car Squadron Mobile Column), Major Langford (156th East African Heavy Anti-Aircraft Artillery Battery) and Lt.-Col. Campbell (Kenya Regiment). The replies from each commander are also in this file.

for kills between units.[35] Eventually, the Cabinet decided upon an inquiry into three particular accusations. Firstly, monetary rewards for killings; secondly, the keeping and exhibition of scoreboards recording official and unofficial kills and other activities in operations; and thirdly, the fostering of a competitive spirit amongst units in regard to kills.[36] In order to proceed impartially, the court was chaired by a Lieutenant-General without previous personal involvement in Kenya, accompanied by Colonel G. Barratt, the Deputy Director of Army Legal Services to guarantee compliance with the rules of procedure, and Colonel G.A. Rimbault, the Deputy Chief of Staff in East Africa Command who had been present from the start of the Emergency.[37]

There are several other reasons for supposing the court would proceed objectively, besides Lieutenant-General McLean's seniority and detachment, in succeeding 'to clean up rather than to cover up', as the Secretary of State for War put it.[38] Firstly, witnesses gave evidence on oath.[39] Secondly, whilst they could face prosecution for perjury, evidence given by a witness could not subsequently be used against them.[40] These measures clearly granted a freedom of expression which might easily have proven troublesome otherwise. Thirdly, investigations were extensive, absorbing information from 147 witnesses over a 12-day period, from every major unit and formation in-theatre. These included staff officers, 12 regiments (even necessitating travelling to Uganda to interview the 4th KAR), a Roman Catholic bishop, and regimental medical officers (RMOs).[41] In terms of ranks, witnesses ranged from Brigadier-General to Private soldier; the most strongly represented group though were in the crucial Major to Second-Lieutenant group, the company and platoon commanders who exercised great influence in the conduct of this decentralized conflict.[42] Eleven National Servicemen took part, as did 14 African Warrant Officers

---

35 PREM 11/696: *Hansard* excerpt, 30 November 1953, p. 770.

36 PREM 11/691: Excerpt from Cabinet Conclusions, minute 2, dated 8 December 1953.

37 WO 32/15834: Telegram from 'Troopers' (War Office) to GHQ East Africa, 9 December 1953.

38 WO 32/15834: Statement by Secretary of State for War Anthony Head to House of Commons, 10 December 1953.

39 WO 32/15834: Telegram Erskine to Adjutant-General, War Office, 12 December 1953.

40 McLean Proceedings, p. 6. This point was explained by Colonel Barratt to Brigadier Orr as embodied in Rule of Procedure 125A, para. G.

41 The full list of units and formations represented is: GHQ Staff officers, 70th Infantry Brigade, 49th Infantry Brigade, 39th Infantry Brigade, 3rd KAR, 4th KAR, 5th KAR, 6th KAR, 7th KAR, 23rd KAR, 26th KAR, 1st Royal Northumberland Fusiliers, 1st Royal Inniskilling Fusiliers, 1st Black Watch, 1st Devonshire Regiment, 1st The Buffs, Kenya Regiment, Medical Officer in Charge of the Civil Native Hospital, Nyeri, Nanyuki C of E Garrison Chaplain, RC Bishop of Nyeri, Head of Consolata Mission, Deputy Assistant Provost Marshall, Deputy Assistant Director Army Legal Services.

42 The complete breakdown is: 3 Brigadier-Generals, 16 Lieutenant-Colonels, 45 Majors, 20 Captains, 5 Lieutenants, 9 Second-Lieutenants, 2 Regimental Sergeant-Majors, 11

from the King's African Rifles. Commanders were required to give their men 48 hours warning and six volunteered to appear as a result.[43] The provision of interpreters, an encouraging attitude towards some nervous officers and the willingness to pursue matters beyond the defined remit further indicated a court resolved to clear the Army's name by thorough, honest examination of uncomfortable issues. Prominent amongst these of course was Captain Griffiths, who himself gave evidence.[44] Reverend J.F. Landregan, padre to 49 Infantry Brigade, stated that everyone from the brigadier downwards was amazed by the revelations and '… we would not tolerate any atrocities of any description against prisoners.'[45] A fellow KAR officer, from the 4th battalion, thought Griffiths got off very lightly,[46] whilst the 1st Black Watch's RMO said his unit generally considered it a unique case.[47]

McLean, Barrat and Rimbault reached the following conclusions. Firstly, they found one instance where two company commanders, with their CO's knowledge, offered 100 shillings to kill the Mau Mau leader Dedan Kimathi, replicating a similar Police reward to which soldiers were not entitled. The Court considered this mistake 'explicable in the circumstances'. When the Devonshire Regiment first entered operations they offered £5 to the first unit to kill a Mau Mau, an act deemed fair and permissible because it happened in a Prohibited Area, where non-combatants were not permitted. In several other cases, officers rewarded their troops with a few beers for working hard – though the Court deemed the practice unproblematic because the reward was not for killing itself. Finally in this category came Griffiths' own admission of offering cash rewards to his troops. Secondly, it transpired Griffiths had exaggerated the existence of scoreboards for his own purposes. They were defined as 'a visual record kept and displayed solely or mainly to foster unhealthy and irresponsible competition in killings between units and sub units.' Statistics were kept for situation reports and assessing military effectiveness, sometimes in restricted situation rooms on charts consolidating official information. At the lower levels, officers either memorized the information or kept it in notebooks or files. The Court found no evidence for 'unofficial kills'. Thirdly, the Court considered the competitive spirit between units, where soldiers might become so concerned to ratchet up higher scores that they began to disregard legal restrictions and kill for its own sake. The vast majority had no idea how many Mau Mau their own or neighbouring units had recently killed; generally, only the adjutant and others participating in administration possessed such information. Some officers knew how many fatalities their unit had inflicted because of continuous engagement in the same area or low, unchanging figures. Soldiers widely recognized the role played by chance

---

Company Sergeant-Majors, 10 Warrant Officer Platoon Commanders, 4 Warrant Officers, 7 Sergeants, 1 Lance-Corporal, 7 Private soldiers.

43 Six commanders recorded having issued the requisite information, although again not every witness was asked whether they had or not.

44 McLean Proceedings, p. 223, p. 268.

45 McLean Proceedings, p. 197.

46 McLean Proceedings, p. 339 (Capt. I. Grahame of Duntrune, 4th KAR).

47 McLean Proceedings, p. 366 (Lt. L.G. Fallows, Black Watch RMO).

in determining which unit killed insurgents, especially when commanders impressed upon them the cooperative nature of many operations, where, for example, one unit might drive insurgents into another unit's path. In any case, units also quantified success with reference to prisoners, arms and ammunition. Wide dispersal throughout the operational areas militated against competition between battalions who rarely made contact. Companies from the same battalion often found themselves isolated for months. On the other hand, the Court found some competitive spirit, but nothing 'beyond the natural rivalry to be found between sub units in all good regiments.'

Finally, McLean investigated the identification of enemy dead by severing the hands, finding issuing fingerprinting kits rendered it obsolete. Also, there were a few isolated cases involving a KAR company (discussed further below) where inhuman practices occurred. These abnormalities remained in the minority, and on the whole the Army's conduct, '... under difficult and arduous circumstances, showed that measure of restraint backed by good discipline which this country has traditionally expected.'[48] While the Court seemed to overwhelmingly vindicate the Army, equally significant from the disciplinary perspective was the instructional role it played. If any doubts remained after the various orders issued by Erskine about his views on how to conduct the campaign, they were dispelled by the Court's norm-reinforcing interaction with representatives from every major unit. Soldiers who stood before the Court took the experience back to share with others in their unit. The whole tone of the questioning and the comments made on evidence given consistently expressed a clear view on what constituted acceptable behaviour. For example, the Court pressed the 39th Infantry Brigade's Brigade Major repeatedly on whether the new fingerprinting kits were adequately distributed to all units.[49] In another case, they pointed out the potential dangers in allowing rivalry between platoons patrolling in the Reserve.[50] Most notably though were the frequent references to the various orders restricting the use of force, shown especially to battalion and company commanders.

The final episode in Erskine's internal campaign aimed at imposing the British Army's traditionally tightly disciplined approach towards counter-insurgencies came on 11 March 1954, when his bete noire was convicted in a second court-martial. Erskine bemoaned how, 'This blasted man Griffiths is giving me more trouble', but new evidence presented the chance to proceed with a murder charge.[51] This time, though, Erskine decided on the lesser counts of grievous bodily harm, and disgraceful conduct of a cruel kind. They concerned a series of related incidents over two days during a patrol lead by Griffiths in June 1953. On 14 June, B Company, 5th KAR patrol collected two Kikuyu prisoners from Embu police station, who were to accompany

---

48 WO 32/15834: 'Summary of Report by the McLean Court of Inquiry into allegations made during the trial of Captain G.S.L. Griffiths, D.L.I., against conduct of the British Security Forces in Kenya', no date.

49 McLean Proceedings, p. 400 (Major W.B. Thomas, 39th Infantry Bde).

50 McLean Proceedings, p. 444 (Major J. Rogers, Devons).

51 IWM Erskine Papers, Erskine in letter to Philip Erskine, dated 2 February 1954.

them on a mission to find Mau Mau. That evening, Griffiths handed his knife to Private Ali Segat, ordering him to threaten one of the prisoners with emasculation. Later on Griffiths instructed him to cut off the prisoner's ear. Afterwards the accused failed to assist the wounded man in any way and left him to suffer in agony. The next morning, Griffiths ordered Segat to pierce the other prisoner's ear with a bayonet and pass through it a long wire so as to lead the man like a dog on the patrol. Although not placed on the charge sheet, one prisoner subsequently died, with substantial evidence that he was murdered and not 'shot whilst trying to escape', as claimed originally.[52]

When the Court handed down a guilty verdict on 11 March, General Erskine's authority and ethical operational concept were both strongly reinforced. Newsinger views this second trial as merely 'a public relations exercise following the public outcry over Griffith's earlier acquittal.'[53] While East Africa Command and the War Office certainly responded to parliamentary and newspaper reactions to the first trial and the McLean Inquiry, the various letters written by Erskine to his family during this period emphasize beyond doubt his immense frustration with 'this blasted man', 'this damned man' whose behaviour was 'absolutely inexcusable and unnecessary' in this 'most revolting and unforgivable case'.[54] Erskine genuinely reviled everything Griffiths had done and stood for, namely the supposition that the Army's conduct depended upon the enemy against whom they were fighting and whether the enemy themselves observed the rules of war. As John Hobson put it in his summing up for the prosecution:

I hope this Court will not accept any such doctrine and will make it plain that so far as the British Army is concerned and its officers it expects its officers to conduct themselves properly and with propriety towards those who are in its custody and against whom they are fighting.[55]

In finding Griffiths guilty, the Court accepted Hobson's opinion, showed command orders on conduct would be implemented and converged with a broader centralizing in the direction of the war effort, seen for example, with the creation of the War Council at the highest level.[56] The discipline imposed not only satisfied ethical-legal concerns in regards to non-combatants, but equally importantly in the British

---

52   WO 71/1221: Proceedings of the General Court Martial of Captain G.S.L. Griffiths, for cruelty.

53   John Newsinger, 'Revolt and Repression in Kenya: The 'Mau Mau' Rebellion, 1952-1960', *Science and Society* 45/2, 1981, p. 179.

54   IWM Erskine Papers, Erskine in letter to Philip Erskine, dated 2 February 1953; Erskine in letter to Philip Erskine, dated 23 February 1953; Erskine in letter to Philip Erskine, 12 January 1954; Erskine in letter to his wife, dated 27 October 1953.

55   WO 71/1221: Second trial of Captain Griffiths, Proceedings, Second Day, p. 10: Prosecuting Counsel Mr J. Hobson.

56   Randall W. Heather, 'Intelligence and Counter-Insurgency in Kenya, 1952-56', *Intelligence and National Security* 5/3, 1990, p. 60.

tradition performed an instrumental role in achieving strategic objectives. Indeed, where other militaries may have resorted to strategies of war crimes in counter-insurgencies precisely because ethics and strategy bifurcated, the British tradition demanded their unity, the resilience of which called for disciplined forces. As we have seen, the command made considerable efforts to secure such discipline.

Indiscipline: 'I Defy Anybody to Bring Some of the Lunatics in this Place Under Proper Control'[57]

As Tony Benn argued shortly after Griffiths' acquittal, simply saying that a thing is prohibited is not the same as saying that it is not the practice.[58] In this observation, Benn astutely recognized the difficulty facing Erskine in a campaign fought at such a low level that a commander could never properly know what was happening. The temptation is to take his orders as an immediately achieved reality; as Anderson argues Erskine 'put an immediate stop to these practices.'[59] This is wishful thinking considering he remained ignorant about many instances until around December 1953, six months later. In this section, it is argued that after an opening eight-month period when the security forces applied force without restraint, indiscriminately, brutally and illegally, for the next nine months East Africa Command waged an only partially successful campaign against its own soldiers to impose a discipline based on new views on how to fight Mau Mau. The campaign was tempered by an awareness that soldiers exerted influence over the command and could only be pushed so far. We analyze the indiscipline within the Army with reference to the weaknesses in the military criminal justice system, Erskine's struggle to impose discipline, the types of crimes which took place, and the local reasons for them.

We already know from Erskine's private papers that when he arrived there had been a lot of indiscriminate shooting going on.[60] The recently released information on the McLean Inquiry confirms this and makes a connection between settler violence and security force collaboration seem very likely. In a letter to the Secretary of State for War in December 1953, Erskine was so concerned about this that he hoped an independent inquiry investigating everything from the beginning of the Emergency would not happen: '... the revelation would be shattering.' The letter continued:

> There is no doubt that in the early days, i.e. from Oct 1952 until last June there was a great deal of indiscriminate shooting by Army and Police. I am quite certain prisoners were beaten to extract information.[61]

---

57 IWM Erskine Papers, Erskine in letter to wife, postmarked 5 August 1954.
58 PREM 11/696: *Hansard* excerpt, 30 November 1953, p. 776.
59 Anderson, *Histories of the Hanged*, p. 258.
60 IWM Erskine Papers, Erskine in letter to wife, dated 28 November 1953.
61 WO 32/15834: Letter from Erskine to Secretary of State for War, 10 December 1953.

This constitutes the first official admission that the security forces, including the Army, participated in widespread murder and torture for an eight-month period during the Kenya Emergency. As there were plans in place beforehand and the military conducted operations during this period, it cannot be argued that any crimes committed by the Army were one-offs. There is not enough direct evidence to prove a policy of war crimes, but given the clear awareness of how certain settlers were evicting people – namely by intimidation and murder – it is probable that the Army were involved as an informal policy. When General Erskine came to Kenya he did not continue the same disciplinary policy already in place. Rather the series of measures discussed in part one were in fact implemented in an attempt to fundamentally change the Army's approach to fighting in Kenya. Erskine's major battle in 1953 was within the Army, not with the Mau Mau, to change the organization's culture away from following the settlers' vision of war fighting and towards what he thought the traditional British method. In a sense then he sought to de-barbarize the conflict.

However, the Army suffered from weak command from the start. Colonel Rimbault was appointed Personal Staff Officer to the Governor at the end of December 1952, but lacking both seniority and staff, he failed to coordinate security force efforts. Hinde was appointed Chief Staff Officer to the Governor on 1 February 1953, but also found he lacked authority and could not coordinate effectively.[62] He was re-designated Director of Operations on 11 April 1953, after the Lari massacre, and started to issue directives but still the campaign lacked direction.[63] This partly explains why Erskine found himself having to spend so much time on centralizing the command system and imposing discipline. Prior to his arrival, Baring failed to take control and thereby actually abetted security force crimes. For example, his 17 April Press Release regarded occasional errors of judgement as inevitable and stated the Government did not welcome 'the general and usually unsubstantiated allegations of brutality and indiscipline that are made both by ill-informed or ill-disposed persons.'[64] Such was the attitude towards complaints. The Army shared this disbelief and some tended to assume any complaints against the Army were by definition lies concocted by Mau Mau supporters in order to have soldiers investigated and thus made ineffective.[65] Apologists even suggested Mau Mau mutilated the bodies of their own dead in order to frame the security forces.[66] Some thought it all merely 'latrine rumours', expressing a general reluctance to accept the possibility that the British Army could engage in widespread brutality.[67] Brigadier Orr remembered having investigated allegations against Mobile Column A early in the Emergency, only to find nothing, although the matter was serious enough for him to issue orders

---

62 Percox, 'British COIN', pp. 70-71.

63 Percox, 'British COIN', p. 73.

64 WO 32/21721: McLean Court of Inquiry Exhibit 29: Press Office Handout, dated 17 April 1953.

65 Julian Paget, *Counter-insurgency Campaigning*, (London, 1967), p. 108.

66 Robert Edgerton, *Mau Mau: An African Crucible*, (London, 1989), p. 158.

67 McLean Proceedings, p. 144 (Major B. Holdsworth, 3rd KAR).

on conduct in response.[68] The allegation regarded the mistreatment of prisoners, and another similar instance came to light in the Inquiry. Here also the evidence proved faulty but the interesting angle is the unit's reaction to investigation:

> ... I would like to say that the impression of most of our subalterns in the company (and I do not think we have got any of the Bolshie breed) is just that everybody feels a little restive that the various hardships of the situation are not fully appreciated in other places.[69]

Despite the Deputy Assistant Provost Marshal's avowed ignorance about any deliberate smear campaign by Mau Mau, others were adamant the campaign existed thanks to Kikuyu cunning.[70] The bias against evidence from Kikuyu witnesses resulted in complacency most tragically evident in relation to rape. According to Clough, the security forces were notorious for this crime, and Elkins documents several cases though unfortunately, if understandably, without identifying the units, places or dates involved.[71] At the Inquiry, Captain Wigram revealed two allegations stood against askaris in the 5th KAR for rape, and the Court completely failed to investigate them, preferring to move onto other topics.[72] Later, when another witness mentioned rape, the Court replied: 'That is not the sort of thing we are concerned with.'[73] The man responsible for Army prosecutions considered rape in the same category as theft.[74] Although Rubin argues that in post-1945 British counter-insurgencies the authorities did not shy away from prosecuting where the prosecuting criteria were met, the evidence on rape suggests that these criteria were not met due to inadequate investigation.[75] This corresponds with the situation in the civil sphere, where Baring protected administration officers from Attorney-General Whyatt's demands for inquiries and prosecutions, so as not to damage morale.[76] Whyatt's Legal Department had a limited investigative capacity with a tiny staff and relied on the information of those whom they were investigating.[77] In theory the Government retained the right to investigate the Army, but in practice the organization stayed almost completely exempt from civilian scrutiny until the mid-1960s, although considering the administration's reluctance to investigate itself there is little reason to believe extended civilian

---

68 McLean Proceedings, p. 13 (Brig. J.H. Orr, 70th Infantry Bde).

69 McLean Proceedings, p. 67 (Capt. R.J. Symonds, 23 KAR).

70 McLean Proceedings, p. 355 (Major C.J. Dawson, DAPM); p. 120 (Major W.A.F. Maynard, 23rd KAR); Heather, 'Intelligence and Counter-Insurgency', p. 70.

71 Marshall S. Clough, *Mau Mau Memoirs: History, Memory, and Politics*, (London, 1998), p. 156; Caroline Elkins, *Britain's Gulag. The Brutal End of Empire in Kenya*, (London, 2005), pp. 247-248, p. 254.

72 McLean Proceedings, p. 261 (Capt. G.F. Wigram, Kenya Regiment).

73 McLean Proceedings, p. 316 (in response to Lt.-Col. D.H. Nott, 4th KAR).

74 McLean Proceedings, p. 350 (Major C.J. Dawson, DAPM).

75 Gerry Rubin, 'Courts Martial from Bad Nenndorf (1948) to Osnabrück (2005)', *Journal of the Royal United Services Institute* May 2005, p. 53.

76 Charles Douglas-Home, *Evelyn Baring: The Last Proconsul*, (London, 1978), p. 250.

77 Berman, *Control and Crisis*, p. 358.

oversight would have made a difference.[78] The same could be said for the military's investigative capacity, comprising until Major Dawson's arrival in November 1953 of one SIB NCO. The regular military police numbers were also low, with considerable problems keeping order amongst leave troops in Nairobi and Mombassa. The number of MPs available never exceeded a total of 60 all ranks.[79]

There are grounds for questioning even the McLean Inquiry's effectiveness as a tool of investigation within the military disciplinary regime. Firstly, as noted the terms of reference restricted any investigation into events prior to 1 June 1953, a situation to which a witness's attention was drawn when thought necessary.[80] Bearing in mind Erskine's statement on indiscriminate shootings and torture in the early period, its exclusion dramatically altered the findings. The same is true to a lesser extent in that several units were missing: the 1st Lancashire Fusiliers, and the still operational East Africa Armoured Car Squadron and the 156th (East African) Heavy Anti-Aircraft Artillery. A mere six volunteer witnesses from the entire Army, none of whom provided any revelations, hardly endorses the Inquiry's truth-seeking credentials. Partly this was due to commanders not giving the stipulated 48 hours notice.[81] Probably the thin red line closed in and soldiers recognized telling tales on men in their own units would cause them an uncomfortable existence in Kenya. For example, surprisingly enough the phrase 'not in my unit' repeated itself as witnesses who admitted hearing rumours remained elusive on the details.[82] There were other flaws in the questioning procedure. One or two witnesses found themselves asked leading questions obviously designed to allow them to prove their virtue.[83] The senior NCOs interviewed, some with immense combat experience, were patronized, only asked simplistic questions and not encouraged to expand in the way that officers sometimes were.[84] On another occasion, Colonel Rimbault expressed his deep admiration for a witness; hardly conducive to an impartial assessment of his role in the campaign, especially as the man in question commanded the notorious 'I Force' from the Kenya Regiment.[85] Having discussed problems affecting investigation, we now analyze some issues of implementation and punishment.

Whilst the position on rape reflected a laid-back attitude towards investigation, that on chopping hands off showed tardiness in implementing policy. As noted, this

---

78 David French, *Military Identities: The Regimental System, the British Army, and the British People, circa 1870-2000*, (Oxford, 2005), p. 181.

79 Lovell-Knight, *Military Police*, p. 161.

80 McLean Proceedings, p. 1 (Lt.-Col. A.D.B. Tree, GHQ); p. 19 (Major I.G. Jessop, 70 Inf. Bde.); p. 154 (Capt. F. Flory, 3rd KAR); p. 211 (Lt.-Col. L.W.B. Evans, 5th KAR).

81 McLean Proceedings, p. 55 (Lt.-Col. J. Cowell Bartlett, 23rd KAR); p. 189 (Lt.-Col. E.H.W. Grimshaw, RIF).

82 McLean Proceedings, p. 97 (Major D.L. Lloyd, RNF); p. 146 (Major R.E. Stockwell, 3rd KAR); p. 160 (Lt.-Col. R.E.T. St.John, RNF); p. 200 (Lt. G.E.M. Stephens, RIF).

83 McLean Proceedings, p. 13 (Brig. J.R.H. Orr, 70 Inf. Bde); p. 155 (Capt. F. Flory, 3rd KAR).

84 For example: McLean Proceedings, p. 177 (CSM J. Bailey, RNF).

85 McLean Proceedings, p. 292 (Major N.M.C. Cooper, Kenya Regiment).

common practice was proscribed on 1 August 1953; but did not disappear as quickly as Erskine wished. The struggle over this issue is symbolic of the command's wider battle to assert disciplinary control over the forces and shows how its authority was limited by those it aspired to command. Whether the practice continued because the hands assumed a trophy value, or to intimidate civilians, or simply became a ritual must remain for the present a speculative point.[86] Ironically, the view that it was British to avoid excess arose from criticism of Belgian outrages in the Congo, including the removal of hands.[87] The Inquiry discovered seven instances since the prohibition came into effect where soldiers mutilated bodies because they did not carry identification kits.[88] A further four people complained about too few fingerprinting kits in their possession, with Lieutenant Colonel Grimshaw pointing out his unit owned none at all.[89] Several units managed by improvising with stamp pads, mud, or pens instead.[90] Nonetheless, it seems that the rather implausible defense of ignorance about the new orders in two cases highlights how easily soldiers could violate command orders with impunity.[91] Indeed, the 23rd KAR's CO stated his brigade (the 70th) and the higher command were in dispute because the fingerprinting kits proved impractical in the field.[92] Further exacerbated by amateurish distribution, the fingerprinting issue displays the ease with which Erskine was ignored, and it is possible this was also the case with less acceptable practices that witnesses declined to comment on at the Inquiry.[93]

Captain Griffiths' acquittal represented a major blow to the command for several reasons. Firstly, the case came to court six months after the offense, meaning Army activities were well hidden from the commander. Secondly, the norm for court-martial convictions stood at around 90 per cent in 1953.[94] Court-martials were not independent legal tribunals but rather part of the chain of command, and therefore expected to

---

86 Edgerton, *Mau Mau: An African Crucible*, p. 167; James Waller, *Becoming Evil. How Ordinary People Commit Genocide and Mass Killing*, (Oxford, 2002), p. 207.

87 Joanna Lewis, "'Daddy Wouldn't Buy Me a Mau Mau': The British Popular Press and the Demoralization of Empire' in Atieno Odhiambo and John Lonsdale (eds), *Mau Mau and Nationhood: Arms, Authority and Narration*, (Oxford, 2003), p. 239.

88 McLean Proceedings, p. 45 (2nd Lieut. Tawney, 6th KAR); p. 123 (Sgt. M.R.N. Tetley, Kenya Regiment); p. 252 (Major J. Gordon, 5th KAR); p. 372 (Major C.M. Moir, Black Watch); p. 413 (Major N.F. Gordon-Wilson, Buffs); p. 433 (Major D.N. Court, Buffs); p. 453 (Lt. J.R. Marshall, Kenya Regiment).

89 McLean Proceedings, p. 156 (WOPC Kitur, 3rd KAR); p. 167 (Major P. Bulman, RNF); p. 176 (2nd Lt. W.D.H. Smith, RNF); p. 187 (Lt.-Col. E.H.W. Grimshaw, RIF).

90 McLean Proceedings, p. 160 (Lt.-Col. R.E.T. St.John, RNF); p. 193 (Major P.M. Slane, RIF); p. 200 (Lt. G.E.M. Stephens, RIF).

91 McLean Proceedings, p. 336 (Capt. I. Grahame of Duntrune, 4th KAR); p. 368 (Major A.O.L. Lithgow, Black Watch).

92 McLean Proceedings, p. 50 (Lt.-Col. J. Cowell Bartlett, 23rd KAR).

93 McLean Proceedings, p. 460 (Major J.N. Holmes, GHQ).

94 French, *Military Identities*, p. 185.

conform to the convening officer's intentions in order to make an example.[95] Thus his acquittal on farcical grounds severely undermined Erskine's authority, and the orders on discipline should be seen as attempting to impose his will on an unruly Army rather than a proactive measure to stop something before it happened. Erskine recognized his weak position in deciding to try Griffiths for a second time on lesser charges, because a second acquittal would have proved disastrous. His determination to succeed is proven by his securing Lord Russell of Liverpool to sit as Judge Advocate, a military lawyer who previously directed all court-martial and war crimes trials in British-occupied Germany. Thanks to this weakness, as Major Clemas, with the 23rd KAR noted, there was: '… a distinction between the formal and the real, that it was necessary to ignore three out of four infringements of discipline and then jump on the fourth.'[96] Possibly the military authorities knew they could use coercion if only a small number of soldiers refused to obey orders.[97] The fact that there were few court martials in Kenya may prove only that the practices the command was trying to stop were too widespread for them to control.

The weaknesses in military prosecution again reflected those in the civilian world, where the courts repeatedly took the side of the security forces and failed to strongly condemn crimes committed by them. For example, judges cared little that confessions were produced under duress, therefore tacitly approving torture.[98] In this sense Erskine's arrival may have marked the beginning of the end of settler dominance in Kenya, but not until after a long fight.[99] This is shown by the light sentences handed down to Keats and Rubens, from the Kenya Regiment and Kenya Police Reserve respectively, for beating Elijah Njeru to death: a collective fine of £150.[100] Erskine could take no further action than seeing them dismissed from the forces.[101] The whole atmosphere is perfectly summed up by the judge in Brian Hayward's November 1953 trial for torturing Mau Mau suspects:

> It is easy to work oneself up into a state of pious horror over these offences, but they must be considered against their background. All the accused were engaged in seeking out inhuman monsters and savages of the lowest order.[102]

The 1955 amnesty for the security forces sought to pass a line under all previous misdeeds and stop any further prosecutions of security forces personnel. Although the settlers were outraged, the amnesty was not really a departure from the norm as it

---

95 French, *Military Identities*, p. 190.

96 Anthony Clayton and David Killingray, *Khaki and Blue: Military and Police in British Colonial Africa*, (Athens, OH, 1989), pp. 239-42.

97 French, *Military Identities*, p. 200.

98 Anderson, *Histories of the Hanged*, p. 101.

99 Anderson, *Histories of the Hanged*, p. 180.

100 Evans, *Law and Disorder*, p. 267.

101 CO 822/471: Telegram from Kenya Governor's Deputy to Secretary of State for the Colonies, dated 12 December 1953.

102 Cited in Elkins, *Britain's Gulag*, p. 83.

had been the de facto policy for years.[103] This late move underscores how minimal the Army's efforts at prosecuting criminals within their own ranks had always been.

The Inquiry revealed a certain degree of resentment within the Army towards the various orders and investigations imposing the new disciplinary regime. For example, some in the KAR felt all the inquiries and investigations were superfluous.[104] Worse still, witnesses stated how 'carping criticisms' and 'legal quibbles' hampered operations and decreased morale.[105] These last two complaints came from the Kenya Regiment, as did others; Captain Franklin thought the court martials confused the troops to the extent that they required instruction on the rules of engagement prior to operations.[106] Captain Guy, the Regiment's 'operational adjutant', believed the Griffiths case and associated press criticism adversely affected them more than other units. Why this was so he did not elucidate; may we speculate the Kenya Regiment's behaviour varied little from Griffiths'?[107] On the other hand, the KAR took note from the court martial and it caused them to exercise extra caution in treating prisoners; the following Inquiry seemed to them a witch hunt in progress.[108] Lieutenant Colonel Windeatt continued the thinly-veiled threats when he said the troops saw the inquiries as 'making an awful lot of fuss out of nothing, they are beginning to wonder what they are sent out here to do.'[109] With this remark the Colonel hit on the crux of the matter in the disciplinary relationship between the command and the units doing the fighting. Erskine pushed hard to make his convictions a reality but there was resistance to new methods from soldiers present for eight months already and holding their own views on how to conduct the campaign.[110] These tensions were evident in the policies on mutilation, and on prisoners. But as we see from the breakdowns in some KAR units, the tensions persisted into beatings, torture and murder.

Combined with already available information, the McLean Inquiry material makes Clayton's argument that the KAR were generally well-disciplined seem rather dubious and unlikely.[111] It should be borne in mind that while significant differences from regular British battalions existed, the KAR fell within the normal military command structure, was modelled on the British Army, officered by seconded British Army men and closely interoperated with them throughout the Emergency.

Regarding beatings, Anderson observes how despite Erskine's efforts, accusations of beatings and mistreatment were a regular feature of the proceedings in the special

---

103   Douglas-Home, *Evelyn Baring*, pp. 260-261; Elkins, *Britain's Gulag*, p. 280.

104   McLean Proceedings, p. 65 (Capt. R.J. Symonds, 23rd KAR).

105   McLean Proceedings, p. 406 (Capt. J.W. Turnbull, Kenya Regiment); p. 296 (Major N.M.C. Cooper, Kenya Regiment).

106   McLean Proceedings, p. 299 (Capt. S.E. Franklin, Kenya Regiment).

107   McLean Proceedings, p. 291 (Capt. R.K. Guy, Kenya Regiment).

108   McLean Proceedings, p. 270 (Rev. C.S. Jerome, KAR Chaplain Nanyuki).

109   McLean Proceedings, p. 438 (Lt.-Col. J.K. Windeatt, Devons).

110   Lonsdale calls Erskine the 'commander during the critical first part of the war'; whereas from the disciplinary perspective he inherited an Army with values alien to his own and struggled to reform them. Lonsdale, 'Mau Maus of the Mind', p. 414.

111   Clayton, *Counter-Insurgency in Kenya*, p. 19.

Emergency courts from June 1953 to the end of 1956.[112] The literature leans towards assuming only the police and Home Guard inflicted beatings on non-combatants rather than the Army.[113] However, the 6th KAR's commanding officer thought his troops '… will give a chap a cuff over the head or give him a bang with the butt if they do not do what they are told.' He considered this only some 'slightly rough handling' rather than deliberate brutality.[114] In 23rd KAR's 'C' Company, Major Rawkins compared the clips round the ear he had seen administered to civilians as 'no beating worse than I got as a boy.'[115] Another officer helpfully defined the common euphemism, 'rough handling': '… I would not call it beating up in the way I visualize a beating up. It was just that they were a little bit over easy with a stick or a rifle butt.'[116] These low level beatings seemed fairly common in December 1953, but less so than at the beginning when they were rife.[117] Indeed, several witnesses failed to see anything wrong in giving a good 'clip across the head' to 'wake a fellow up.'[118] The Kenya Regiment Afrikaners achieved special notoriety for their 'rough and tough' outlook.[119] One Kenya Regiment soldier working with the Black Watch, apt to knocking civilians about, was reprimanded and instructed to conform to Black Watch standards whilst with them.[120] On the other hand, Captain Duntrune freely confessed having beaten suspects on the bottom with a stick – 'That I have done quite frequently.' In fact, he went on, '… other Europeans have done exactly the same and it was one of the best ways of getting information, but so far as torturing is concerned I know nothing of it.'[121] Of course the distinction between beating and torture is far from axiomatic, but the Army seemed to differentiate them and generally considered a little beating judiciously applied more helpful than harmful.

Another productive method for obtaining information during interrogations rested in the art of terrifyingly convincing intimidation. Second Lieutenant Muir from the Kenya Regiment recalled interrogators grabbing the victim by the scruff of their neck when threatening them.[122] 3rd KAR Major Topham banned torture, but allowed frightening a prisoner immediately after capture in order to obtain information: 'I myself would adopt any means of frightening him which did not amount to, speaking quite honestly, torturing or maiming or in any way really impairing his possibility

---

112   David M. Anderson, 'The Battle of Dandora Swamp: Reconstructing the Mau Mau Land Freedom Army October 1954' in Odhiambo and Lonsdale, eds, *Mau Mau and Nationhood*, p. 164.
113   Clayton, *Counter-Insurgency in Kenya*, p. 13.
114   McLean Proceedings, p. 27 (Lt.-Col. R.C. Glanville, 6th KAR).
115   McLean Proceedings, p. 57 (Major H.F. Rawkins, 23rd KAR).
116   McLean Proceedings, p. 65 (Capt. R.J. Symonds, 23rd KAR).
117   McLean Proceedings, p. 258 (2nd Lt. R.E. Ginner, 5th KAR).
118   McLean Proceedings, p. 261 (Capt. G.F. Wigram, Kenya Regiment); p. 291 (Capt. R.K. Guy, Kenya Regiment); p. 72 (2nd Lt. J.Y. Ellis, 23rd KAR)
119   McLean Proceedings, p. 296 (Major N.M.C. Cooper, Kenya Regiment).
120   McLean Proceedings, p. 360 (Lt.-Col. D. McN. C. Rose, Black Watch).
121   McLean Proceedings, p. 337 (Capt. I Grahame of Duntrune, 4th KAR).
122   McLean Proceedings, p. 115 (2nd Lt. I.K. Muir, Kenya Regiment).

of life.'[123] Captain Russell's account of how he frightened a prisoner without hurting him is illuminating:

> I quite deliberately took my pistol, put the chamber of the pistol against his left ear and knowing that the barrel protruded behind the man's head I fired one shot into a sack lying on the ground behind. The man was very frightened. Further questioning elicited no information at all and I decided then that the man knew nothing whatsoever about this gang.[124]

Torture from this point on became a short step, even more so when intelligence was required. Throup argues that the demand for intelligence in the early stages, accentuated by Mau Mau's effective intimidation campaign, could not be met.[125] The beatings, intimidation and torture instigated by British forces seem to have come in response to this drought, with, for example, Kenya Regiment commander Guy Campbell conceding soldiers used torture to extract information whilst denying knowing about any specific instances.[126] The Inquiry unveiled two other instances where torture happened besides those carried out by Griffiths; the full records for both of these are still under Government lock and key. The first concerned Captain Hilbourne-Clarke from the 7th KAR, who allegedly tortured a prisoner by burning him with cigarettes over several days. As far as is known, he did not face prosecution. In another known incident, called the '10 Somalis case', men in the 23rd KAR under WOPC Hussein were alleged to have taken ten Home Guard members into the forest and executed them. The new McLean evidence shows that in fact two groups of ten were murdered over a two-night period rather than on one day as previously thought. Weaknesses in the investigation mean the situation remains murky. These examples seem to fit in with existing knowledge, for example on KAR involvement in retaliatory massacres after the Mau Mau bloodbath at Lari in March 1953.[127] The Inquiry also presents some initial clues as to why these KAR units fell apart. Firstly, and most significantly, the officers were badly overstretched with not enough platoon commanders, and those who served lacked experience. Secondly, the battalions fought the insurgency from beginning to end and spent longer in the field than any units from Britain. Thirdly, soldiers had to adapt their behaviour when the new commander, Erskine, came into power, and this organizational change naturally took time, whereas new units coming from Britain immediately applied the C-in-C's conception of how to fight. Lastly, the King's African Rifles units were re-enforced by, and most susceptible to, the influence from Kenya Regiment NCOs and officers, members of the settler community with a generally virulent hatred for Mau Mau and sometimes all Kikuyu. British units taking on Kenya Regiment personnel were

---

123   McLean Proceedings, p. 136, p. 138 (Major R.N. Topham, 3rd KAR).
124   McLean Proceedings, p. 278 (Capt. H.C. Russell, 7th KAR).
125   Throup, 'Crime, Politics and the police', p. 144.
126   McLean Proceedings, p. 284 (Lt.-Col. G.T.H. Campbell, Kenya Regiment).
127   Edgerton, *Mau Mau: An African Crucible*, p. 80.

able to overwhelm their prejudices and keep them in good order, whereas the KAR's weaknesses prevented this from being the case.

When Erskine took over the KAR were fighting the Kenya Regiment's sort of dirty war. Erskine could claw back the KAR after a series of long disciplinary measures, but the Kenya Regiment were uncontrollable and in any case invaluable so long as their atrocities were not severe enough to turn the entire population into insurgents. The Army managed to rely upon Kenya Regiment crimes, and crucially for intelligence reasons, the even worse Home Guard, whilst simultaneously distancing themselves and denying involvement.

'Turning the Nelson Blind Eye?'

As we have seen, it is possible to view barbarization not only from the perspectives of environmental factors, retaliation, and indoctrination, but also from a disciplinary angle. This approach is best applied in circumstances where there is no direct war crimes policy, such as in Kenya and other counter-insurgencies where a strategic objective is to win the population over from the insurgents. The Army operated under the rule of law at all times and sustained measures were taken, both preventative and punitive, to ensure soldiers operated within its boundaries. Orders were issued, court-martials conducted and a comprehensive, high-level inquiry carried out. However, these efforts should not be seen as evidence that the Army fully succeeded in implementing its intended minimum force strategy without compromise. As the evidence from the McLean Inquiry now highlights, from June 1953 to March 1954 General Erskine spent much effort battling with his own forces to impose his vision of how to fight the campaign on troops who had already developed another approach. There is now official recognition that this meant widespread and indiscriminate beatings and shootings of the Kikuyu population. The command never fully succeeded in eliminating the barbarous behaviour of certain sections of the security forces but instead managed them by relying upon peripheral units, such as the Kenya Regiment and Home Guard, to do the dirty work and keep the British conscience clear. Further research is desperately needed into the initial stages of the Emergency especially to discover in particular the connections between the Army and settler vigilante groups in addition to further work on the Kenya Regiment and Home Guard.

In terms of the relevance of the Kenya experience for current warfare, it is extremely pertinent in a time when insurgencies are common forms of war and the counter-insurgent forces must try and fight with some restraint in order to win. What this case also highlights, however, is that no matter how professional the Army or powerful its command, war crimes will probably be inevitable and the most that can be done is to manage and reduce the number of incidences. Primary amongst any such measures would be ensuring units were given adequate support in the field, rest and leave; and that the fine line between motivating the troops to fight was not crossed into dehumanizing the enemy to the point where killing them with impunity seemed a natural thing to do. The Kenya case also especially highlights the need for effective investigation and prosecution and that commanders monitor their troops' behaviour even when not directly controlling them. In these days of multinational and all-arms war fighting, such as that in Afghanistan and Iraq, the temptation again arises to rely upon local or alliance forces which may use barbarous methods. In these circumstances 'turning the Nelson blind eye' is a mistake in both moral and strategic terms.

Chapter 5

Beyond Terror and Insurgency: The LRA's Dirty War in Northern Uganda

Anthony Vinci

The Lord's Resistance Army (LRA) is arguably the most feared armed group in the world. It regularly commits violent atrocities such as abducting and mutilating young children and in so doing, the group has systematically created a pervasive climate of fear which is unrivalled in modern warfare. So profound is the level of dread that the children in northern Ugandan villages walk miles to major towns to sleep in the city centre rather than in their own homes and even well-armed soldiers refuse to travel alone. Yet, the LRA's motives are so obscure that it is hard to identify rational explanations for why they are terrorizing the population of northern Uganda.

The LRA is certainly not the only armed group to terrorize society – what separates it is the degree to which it uses fear. Although it is often considered only a side effect of conventional fighting, in asymmetric warfare fear is more methodically used. Traditional insurgents use fear, in the form of surprise attacks, in order to take advantage of its force multiplier effects. Terrorist groups use it strategically to achieve political goals. The LRA, however, does not seem to fit into either of these categories. Its use of fear is so exaggerated and 'senseless' that it is difficult to attribute it to tactical or strategic reasoning.

This chapter will examine the ways in which the LRA has created a climate of fear and will attempt to provide an explanation for why it has done so. It will first look at the history of the LRA and its force structure in order to illustrate how it has been able to so effectively use fear. The chapter will then illustrate the specific tactics that the LRA has used to create fear, including: maximizing threat, surprise, and creating an atmosphere of unpredictability. The LRA will be compared to insurgencies and terrorist groups in order to explain how it is different than either category of armed group. The chapter, finally, posits that the LRA has strategically used fear to fight a

  * I am grateful for comments on earlier drafts from Christopher Coker, Rune Henriksen, Jennifer Rumbach, Hannah Vaughan-Lee, participants at the Barbarization of Warfare conference, and the North-South Seminar at the London School of Economics. I would also like to thank Tony Odiya and Paul Kalenzi for their friendship and assistance in Uganda.

'dirty war'. The consequence of the LRA's unique strategy is that neither counter-insurgency nor counter-terror techniques will provide an adequate response; instead new and specialized tactical and strategic responses must be developed to combat it.

History of the Lord's Resistance Army

The roots of the conflict in northern Uganda go back, at a minimum, to the colonial period. However, the origin of the LRA as an organization occurred after Yoweri Museveni and his National Resistance Movement (NRM) took control of Uganda in 1986. At this time, multiple rebel movements arose to combat the takeover.

The Uganda People's Democratic Army (UPDA) grew out of the defunct Ugandan National Liberation Army (UNLA) and various other anti-NRM forces in an attempt to capture power from Museveni. The UPDA had credibility within the Acholi community because of its anti-Museveni stance. This allowed it to recruit soldiers with relative ease and, therefore, use both conventional and guerrilla tactics against the NRM and its military wing, the National Resistance Army (NRA). The UPDA also knew and used terror tactics, a point of significance to note in regards to the future of the LRA.[1]

Another group, the HSM, founded by Alice Auma (Lakwena) in 1987, also attempted to battle the NRA. Alice, who claims to be possessed by a World War I era Italian named Lakwena, had significant Acholi support and volunteers. Also, as the UPDA lost some ground against the NRA, Alice recruited some UPDA units. The HSM used initiation rituals based on Christianity and local beliefs to create a unique religious-military organization. Alice developed a new bread of military tactics, the so-called 'Holy Spirit Tactics'.[2] Through initiation, purification, and ritual, members of HSM were led to believe that they were invulnerable and that other magical benefits were theirs to use, including the ability to turn stones into grenades and bees into allies. These spiritual tactics were combined with conventional tactics and an organizational structure modelled on the British colonial format.[3] While Alice was able to win a major battle in 1986, she and the HSM were ultimately defeated in November 1987. Alice's father, Severino Likoya Kiberu, attempted to reform the HSM after its defeat. While Alice had been able to rely on charisma and a willing population, Severino was forced to turn to violent means of finding support. To illustrate the point, Doom and Vlassenroot report that one of Severino's names was *otong-tong,* which translates as 'one who chops victims to pieces'.[4] As the LRA would also come to do because of its unpopularity, Severino was forced to rely more heavily on the use of child soldiers.

1   Frank Van Acker, 'Uganda and the Lord's Resistance Army: The New Order No One Ordered', *African Affairs*, 103, 412, 2004.

2   Heike Behrend (tr. by Mitch Cohen), *Alice Lakwena & the Holy Spirits: War in Northern Uganda, 1985-97,* (Oxford, 1999), p. 107.

3   Behrend, *Alice Lakwena & the Holy Spirits*, p. 110.

4   Ruddy Doom and Koen Vlassenroot, 'Kony's Message: A New Koine? The Lord's Resistance Army in Northern Uganda', *African Affairs*, 98, 390, 1999.

In 1987, out of remnants of existent anti-NRA movements, Joseph Kony, a high-school drop-out and former alter boy from Gulu district, founded what would become known as the LRA. The LRA was formed when the more traditional rebel army of the UPDA was combined with the decidedly unorthodox HSM. Kony began his war by taking control of a unit of the UPDA. The aviator glasses-wearing, dread-locked Kony, who is said to be a charismatic leader, was able to attract volunteers from both the UPDA and the HSM. After HSM's defeat and the signing of a peace accord by the UPDA, there were no other effective military groups to represent the Acholi against the NRM/A. Given this, many veterans from both movements joined Kony. In particular, it is reported that the most brutal officers were attracted to Kony's movement because they felt that they had nowhere to go back to, due to the atrocities they had committed.[5]

At first Kony's movement seemed to be a continuation of the HSM. He was 'the bearer of an apocalyptic vision, a mouthpiece of a widely accepted view that the Acholi people [were] on the verge of genocide.'[6] Like Alice, Kony claimed to be possessed by spirits, including a Sudanese, Chinese, American, and former minister of Ida Amin. He also instituted cleansing processes, initiation rites, and practiced mystical acts which, amongst other things, allowed him to make his followers invincible by ritually armouring them with *malailka*, the Swahili word for angel.[7] However, while the original HSM had strict moral rules of behaviour, which helped increase local popularity, Kony's movement had far looser rules, and never achieved the popularity of either the HSM or the UPDA.

The holy spirit tactics were replaced with guerrilla tactics when remnants of the UPDA, led by Odong Latek, joined Kony in 1988. Other commanders, including Tabuley[8] and Vincent Otti (who is currently considered second-in-command) also came over to the LRA voluntarily and have continued to contribute to the guerrilla warfare tactics used by Kony. These commanders brought with them guerrilla tactics and the military experience of using terror as a strategy.[9] To this day, the LRA can be seen as a combination of these two movements, combing the brutal guerrilla tactics of the UPDA as well as, particularly in its training and command and control system, the spiritual tactics of the HSM.[10]

---

5   Lawrence E. Cline, 'Spirits and the Cross: Religiously Based Violent Movements in Uganda', *Small Wars and Insurgencies*, 14, 2, 2004.

6   Doom and Vlassenroot, 'Kony's Message: A New Koine? The Lord's Resistance Army in Northern Uganda', p. 22.

7   Behrend, *Alice Lakwena & the Holy Spirits*, p. 114.

8   Who reportedly died in 2003.

9   Van Acker, 'Uganda and the Lord's Resistance Army: The New Order No One Ordered', p. 348.

10 The evolution of the LRA can be seen in its name changes. The initial movement was called the Holy Spirit Movement II – a blatant attempt at following the HSM. Then it became the Lord's Salvation Army, presumably to distance itself from the HSM. When many former UPDA fighters joined and the army became more of a guerilla force and held less emphasis on the spiritual side, it became the United Democratic Christian Force. Finally, with the death of

The period of the early- to mid-1990s is considered to be a major turning point in LRA strategy for two reasons.[11] It was at this point that the Khartoum Government began supplying Kony's army with new weapons and equipment and allowed the LRA to establish bases in Sudan.[12] In addition, Kony reportedly began to feel that the Acholi people had betrayed him. In particular, he felt this because of the 'bow and arrow' civil defence militias made up of fellow Acholis, set up by the Government, and by the failure of a peace process that had been championed by Betty Bigombe in 1994.[13]

After this period, the LRA directed much more brutal violence at the Acholi people. Major massacres occurred at Atiak in 1995, the Karuma and Acholpi camps in 1996, and Lokung-Palabek in 1997. Similar massacres continue to the present day.

This behaviour is not uncommon in insurgencies, for guerrilla movements that no longer feel they need the local population to survive may turn against it. This occurred during the Greek civil war, where the Communist guerrillas had access to bases in countries north of Greece.[14] The possession of these bases meant the guerrillas 'felt free to express their profound, even murderous, contempt for the Greek peasantry among whom they operated.'[15]

With the loss of what little popular support there was, the LRA instituted an increase in the number of child abductions. From the beginning, the LRA had considerably less support than either the UPDA or HSM, and therefore experienced difficulties in finding new recruits. This had detrimental effects on its ability to conduct a war against the NRA. Although the LRA had committed atrocities, as the UPDA before them, and used kidnapping for recruitment, as Severino's HSM had done, the period after the failure of the 1994 peace process fundamentally changed the way in which the LRA operated. Starting in the mid-1990s, many of the largest kidnappings occurred, such as the 1996 Aboke kidnapping, and they became a much more regular occurrence in the north.

Since then, the practice has only been further reinforced. The adoption of the Amnesty Accord in 2000 threatened Kony's ability to retain personnel who could

---

Latek and return to a more spiritual basis, Kony's army became the Lord's Resistance Army in 1992.

11 Doom and Vlassenroot, 'Kony's Message: A New Koine? The Lord's Resistance Army in Northern Uganda'.

12 This aligning of an Islamic Fundamentalist government with a Christian Fundamentalist rebel group seems surprising at first glance. However, it makes sense from a strategic perspective. The Sudanese support for the LRA was allegedly a retaliatory move against Museveni because he backed the Sudanese People's Liberation Army (SPLA) – the primary rebel movement in southern Sudan. From the LRA's perspective the alliance made sense because it allowed it to strategically gain strength to fight its own war.

13 Doom and Vlassenroot, 'Kony's Message: A New Koine? The Lord's Resistance Army in Northern Uganda'.

14 Anthony James Joes, *Resisting Rebellion: The History and Politics of Counterinsurgency*, (Lexington, KY, 2004).

15 Joes, *Resisting Rebellion: The History and Politics of Counterinsurgency*, p. 19.

now opt to return to society.[16] Kony has since lowered the age of desirable soldiers, taking higher percentages of those aged nine or ten and letting the mid-teenagers, who used to be the most desirable to the LRA, go.[17] The assumption is that the lower age of combatants makes it easier for the LRA to keep abductees from running away.

The alliance with Sudan provided significant military benefit. The LRA created rear bases in friendly Sudan which housed large numbers of LRA fighters, their wives, and other personnel. Excursions into northern Uganda were used to attack soft targets and loot goods. At the same time, forces were used to covertly fight the SPLA in southern Sudan. The LRA also used Sudan as an excellent source for weapons. At some points in the conflict, the LRA has even had better weapons than the Ugandan People's Defence Force (UPDF) (formerly known as the NRA) troops.

Operation Iron Fist in 2002 and Iron Fist II in 2004 have also contributed to the strategic and operational evolution of the LRA. These operations allowed the Ugandan UPDF to follow up and attack the LRA in southern Sudan after an agreement Uganda made with the Sudanese Government. The assaults significantly weakened the LRA's fighting ability and, reportedly, forced it to leave its Sudanese bases. The UPDF's use of helicopter gunships has been particularly effective at disrupting the LRA by making it much more difficult for them to hide. The effect of these operations has been to uproot the LRA and make them even more mobile. The events have combined to allow the LRA to revolutionize its tactics and have contributed to its evolution into the organization that it is today.

LRA Organization

The picture of the LRA that we are left with now is one of an organization consisting of 500 to 1,000 'hardcore fighters', that is adults, probably left over from the HSM and UPDA.[18] Combined with these soldiers are a fluctuating number of abductees, usually estimated to be approximately 3,000 at any given time.

Kony is the ideological leader and senior general of the LRA. Below Kony are 'Brigadiers'. These include Vincent Otti, in charge of military strategy; whoever replaced Kenneth Banya, the Soviet trained tactician of the LRA, who was captured in July 2004; and whoever replaced Samuel Kolo, who was in charge of public relations until his surrender in February 2005.[19]

---

16 The Amnesty Accord gave blanket amnesty to all LRA fighters who returned from the bush. (Refugee Law Project, 'Behind the Violence: Causes, Consequences, and the Search for Solutions to the War in northern Uganda', Working Paper No. 11, (Kampala, 2004), p. 6.)

17 Human Rights Watch, 'Abducted and Abused: Renewed Conflict in Northern Uganda', (London, 2002), p. 21.

18 The exact number is unknown; this number was gleaned from various interview and analyst estimations. Some estimates put the number as low as 200 and others upwards of 2,000.

19 IRIN News 'LRA "Colonel" returns to Kampala', (4 February 2002).

Beyond this core group that has volunteered for the LRA, there are the abductees and those born into the LRA. The LRA no longer takes volunteers; therefore new recruits are always abductees. The abductees are usually young men and women between the ages of eight and 18, though children abducted at five or six are not uncommon. Both sexes are used as porters and soldiers, though women usually become a 'wife' of one of the commanders at some point and then are no longer allowed to serve as soldiers. Finally, there are those who are born into the LRA and have grown up into soldiers. It is said that they are the most brutal, and therefore the most dangerous soldiers, since they do not have any ethical grounding outside of the organization.[20]

Force Structure[21]

The LRA is composed of four brigades – Control Alter (also known as 'Trinkle') which includes the overall leadership, Sinia, Stockree, and Giiva.[22] Each brigade is estimated to have between 300 and 800 members and consists of three battalions. Battalions have their own commanders and are divided into sub-units of around 15 or 20 persons, led by field commanders. Field units tend to be the operational units of the LRA. Field commanders are likely to be abductees who have been with the LRA for a long time and have proven their loyalty.[23] Each unit also has a mid-level 'religious officer' who is responsible for prayer, fasting, and other spiritual duties.[24] During attacks and raids the groups may further split into units of two or three for maximum dispersion. The small unit size of the LRA and their operational independence, similar to conventional armies' special forces units, allows for comparable small-scale, fast-paced, and independent operations.

Operations and Equipment

The LRA units are all self-sufficient and can survive in the bush for long periods of time. Wives and porters are brought along. The units loot their food and all other material needs.[25] Normally, they do not carry weapons heavier than Rocket Propelled Grenades (RPG) and bring only the food necessary for survival. If the unit loots a lot of food or other goods, it will temporarily abduct adults or children to carry it. These units are extremely light and fast. They can reportedly make it from Lira to Gulu, a distance of around 50 miles through heavy bush, in around five hours on

---

20 Interview with UN Access Coordinator, Gulu, 17 January 2005.

21 Note that the force structure, operations and equipment described here are all idealized forms. In practice, the LRA tends to be much more flexible, disorganized, and *ad hoc*.

22 International Crisis Group, 'Northern Uganda: Understanding and Solving the Conflict', (Nairobi/Brussels, 2004).

23 For instance, to become a field commander one must be shot at least once..

24 Interview with UN source, Gulu, 18 January 2005. Confirmed in other interviews.

25 They are reportedly particularly fond of sugar cane which is high in energy and easily transportable. (Interview with local resident, Gulu, 18 January 2005).

foot.[26] If at any time the unit feels that it is being slowed down by abducted porters or wives, it will kill or release them. The units are closely coordinated by radio, or more commonly, cellular phones.[27] Even the smallest units are reported to have some sort of communications device. Though they are not in constant communication, as illustrated by left over attacks after Kony called ceasefires, the units are able to coordinate their movements enough to assure that there is little unwarranted overlap. Furthermore, this allows them to rejoin other units in order to head back to Sudan or combine into a larger attack force. In general, field commanders seem to be given a high degree of control over their unit's activities and the self-sustainable nature of the units means that they can carry out the field commander's instructions indefinitely. The LRA's light, small, fast tactics do not require heavy weapons. Units generally have RPGs for ambushes as well as assault rifles, usually the ubiquitous AK-47s. However, the LRA is notoriously frugal with its rifles and only select members of the unit, usually just commanders, have one.[28] The rest of the soldiers use machetes, axes, or even just a club.[29]

Fear in War

Out of this force structure and particular historical formation from the HSM and UPDA, the LRA has developed a unique strategic use of fear in war. Indeed, it has come to systematically construct a climate of fear in northern Uganda. It is not that fear in war is new. What is unique about the LRA are the ways and extent to which it has used fear strategically. Fear in war is a prevalent, if not the prevalent, feature of war. Major Gregory A. Daddis, US Army, defines fear as 'a physical and emotional response to a perceived threat or danger.'[30] We find so much fear in war because 'combat is about wounding and death and produces much anxiety over anticipated physical harm.'[31]

---

26 Interview with UN Access Advisor, Gulu, 14 January 2005. Admittedly this is an almost unbelievable pace, but it is conceivable. Even if untrue, the pace gives an idea of the quasi-mythical status of the LRA.

27 The LRA reportedly stole radios from the Catholic Church, which still is able to track their transmissions (though they are encoded). Often the units will have to walk to known areas of the bush where they can receive cell phone service and occasionally must send someone to buy phone credit from a town. There are numerous stories of dirty, ragged men walking out of the bush and buying a million shillings worth of phone credit. (Gathered from various interviews conducted in January 2005).

28 For instance, the penalty for escape is execution, but the penalty for escape with a rifle is execution and the execution of one's family.

29 The weapons also have the benefit of being cheap, light, not needing ammunition, and exact in that victims can be kept alive.

30 Gregory A. Daddis, 'Understanding Fear's Effects on Unit Effectiveness', *Military Review*, July-August, 2004, p. 22.

31 Daddis, 'Understanding Fear's Effects on Unit Effectiveness', p. 24.

Fear can be deconstructed into three constituent aspects. These are: *fear*, which is focused on a definite object; *fright*, in which a person is surprised by a dangerous situation; and *anxiety*, which is the state of mind caused by preparation for danger.[32] Warfare runs the gambit of fear. However, in conventional warfare, fear is usually seen as a side effect of war, not something that is systematically created. Conversely, in asymmetrical forms of warfare, fear is more consciously used.

More specifically, insurgents and terrorists are known to commonly use fear for tactical and strategic advantage. Insurgencies use fear as a force multiplier against the military organization they rival. Since their numbers are small, guerrilla fighters use fear to present themselves as a more powerful organization than they really are. At the same time, insurgents commonly use fright, through tactical surprise, to further multiply their forces. Terrorists, on the other hand, expressively use fear as a way to put political pressure on a government. In their case, anxiety is methodically created through seemingly random, violent attacks targeted at unprepared civilians. The LRA combines the use of fear found in both insurgencies and terrorist groups. Like an insurgency, the LRA has used surprise and the creation of a larger perceived threat to multiply its forces. Like a terrorist group, it has created an atmosphere of anxiety. However, in its war, the LRA has maximized the entire typology of fear, unlike either insurgencies or terrorist groups. The following section will illustrate the specific ways in which it has done so.

Forms of Atrocity

The most obvious way in which the LRA has strategically used fear has been to maximize its perceived threat. This has been accomplished through a campaign of conspicuous atrocity in the form of mutilation and civilian abduction, as well as fabricating the appearance that the group is fearless and omnipotent.

Mutilation

Committing atrocities has long been a favourite method of armies to frighten their adversaries. In a study of infantry veterans from the Pacific theatre in World War II, it was noted that 'observations of high numbers of casualties in one's group, coupled with the death of one or more friends, were very important determinants of combat fear, but *almost as powerful a contributor was witnessing the enemy carrying out atrocities.*'[33] In other words, beyond actually attacking your adversary, the best way to frighten him is to commit an atrocity that he observes.

The LRA threat is maximized through a history of conspicuous atrocities. Standard mutilation practices involve severing off lips, ears, noses, fingers, and hands. Other methods include sewing eyes shut and padlocking lips together. A

---

32 Sigmund Freud (C. J. M. Hubback translator), *Beyond the Pleasure Principle*, (New York; 1922) framework illustrated in Christopher Coker, *Humane Warfare*, (London, 2001).

33 Stanley J. Rachman, *Fear and Courage*, (San Francisco, 1998), p. 71 (italics added).

typical story, reported by Father Carlos Rodriquez, is of a boy of 17 who has had his ears, lips and fingers cut off by the rebels. They wrapped his ears in a letter which they put into his pocket. The letter read, '[w]e shall do to you what we have done to him' – referring to those who wanted to join the local defence forces.[34] Victims are often forced to 'consent' to the mutilation by being silent throughout the process on pain of death. Even their choice of weapons, axes and machetes, serve as a visible sign of threat.

Such LRA brutality is highly visible and symbolic,[35] allowing it to use mutilation as a method of communication and control over the population. Ears and lips are cut off as a signal to beware of informing on the LRA. Bicycle riders have their legs cut off because bicycles, a major mode of transportation, also bring communication. Rape is often public, as a way to humiliate both the victim and his or her family members.

These signals allow the LRA to institute control over the population. For instance, if the LRA wants to limit communication between populated areas, it must control the number of bicycles travelling between population centres. A checkpoint system would be impossible given the size of their forces relative to the distances and populations involved. By visibly mutilating bicyclists, the LRA provides an effective reminder of the unacceptability of cycling and in this way can efficiently limit communication with a minimum use of manpower.

Abduction

Another form of atrocity is the widespread abduction carried out by the LRA. The LRA is known for its reliance on abducted soldiers, sex slaves, and labourers to fill its ranks. Human Rights Watch reports that more than 20,000 children have been abducted since 1990.[36] The horrible future that awaits those who have been abducted, in some ways worse than death, makes it a powerful threat. The typical abduction is extremely brutal. Intimidation and beatings are followed by being forced to carry loads for long distances. A process of initiation through traumatization then takes place, in which the children will be ordered to torture and kill fellow abductees. A standard method is to put victims in the middle of a circle and have the rest beat them to death. Then there is 'registration', in which each abductee receives 50 or more lashes. This is followed by months of hard labour, regular beatings, food deprivation, and summary execution for disobeying orders or attempting to escape.

The threat of abduction has created a climate of fear throughout northern Uganda. It has contributed to the internally displaced persons (IDP) camp siege mentality that dominates the north, where 1.4 million civilians are confined to government-

---

34 Father Carlos Rodriguez, 'War in Acholi. What can we do?', (Akron, PA, 2003), p. 1.

35 Civil Society Organizations for Peace in Northern Uganda (CSOPNU) 'Nowhere to Hide: Humanitarian Protection Threats in Northern Uganda', (Kampala, 2004).

36 Human Rights Watch, 'Abducted and Abused: Renewed Conflict in Northern Uganda', p. 20.

protected camps.[37] The situation is so bad that the previously unheard of phenomenon of 'night commuting' has emerged, in which children from villages surrounding urban centres stay within the confines of the city centre in order to escape the threat of abduction.[38] A convenient side effect of the LRA 'initiation' process is that it makes abductees easier to control, extremely violent, and fear returning to society. The LRA initiation process is used to traumatize the children into a psychological state where control becomes easier, as evidenced by the LRA's ability to use child soldiers without drugs, the near universal method for 'priming' child soldiers for combat. This process also makes the children extremely violent themselves, with stories of sociopathic returnees killing siblings because they 'would not be quiet'.[39] At the same time, abductees become less likely to escape because they fear that they will be blamed by their families and villages for committing atrocities.[40]

Fearlessness and Omnipotence

The LRA combines this perceived threat with a constructed perception of its own fearlessness and omnipotence. Most strikingly, abducted soldiers are trained not to take cover, but to march forward with rifle in hand through enemy fire. This training, derived from the HSM holy spirit tactics, involves rituals of invulnerability in which shea butter and holy water is spread on the fighters before combat as well as the threat of punishment for those who fear attack. The tactics are combined with the infamous ability of child soldiers to inflict fear on their adversaries. This ability arises from their own fearlessness in combat and complete disregard for human life.[41] Kony has improved this advantage by brainwashing the child soldiers – using the aforementioned process of initiation through traumatization.

Kony's alleged spiritual powers are used to construct an image of omnipotence. He has created for himself a cult-like belief in his own spiritual powers. Spirits are a common feature of Acholi culture and this makes the belief in Kony's own spiritual

---

37 United States Agency for International Development (USAID), 'Uganda Complex Emergency Situation Report #2 (FY 2004)', (Washington, 2004) based on World Food Program data.

38 Night commuting even occurs in the IDP camps where children who live in the periphery of the camp are now coming to sleep in the centre. (Interview with UN source, Gulu, 18 January 2005).

39 Interview with local aid worker, Gulu, 10 January 2005.

40 Alex de Waal notes a similar approach taken by RENAMO: 'RENAMO training socialized its recruits into extreme violence. They were brutalized, and compelled to participate in acts of atrocity such as killing defenseless people, or other recruits who failed to obey orders. Once they had become stained with blood, as they saw it, the soldiers felt dehumanized, unfit for normal human society, and worthy only of the company of their new peer group – other killers.' (Alex De Waal 'Contemporary Warfare in Africa' in *Restructuring the Global Military Sector, Volume 1: New Wars*, edited by Mary Kaldor and Basker Vashee, (London, 1997), p. 318).

41 See for instance, De Waal 'Contemporary Warfare in Africa'.

power, especially amongst children, common. In some cases the abducted children simply believe he is God.[42] This reputation for omnipotence and magic allows Kony to keep control over his own forces and to, at the very least, bring doubt to some of the civilians and even government soldiers of the north. For example, a government official admits, 'it is true that some UPDF troops believe that Kony may have spiritual powers. Some lack of vigour in pursuing rebels can be ascribed to this.'[43]

Fright

The LRA also takes full advantage of the element of surprise. As Anthony James Joes has put it, 'surprise is a true force multiplier, compensating for inferiority in numbers.'[44] Traditional guerrilla strategy involves not just attacking the flanks and rear of an army, but generally catching it off guard. In doing this, the guerrilla army inflicts casualties and creates anxiety within the ranks of the (conventional) army it is fighting against.[45]

The primary tactics of the LRA are raids and ambushes on soft targets. Both forms of attack are helped by the savannah environment of the north, particularly its high grasses, which obscure the view into the bush from the road. This gives the LRA cover when moving and carrying out an attack. The LRA's ubiquitous roadside ambushes may be on public and private transportation, including buses, taxis, cars, trucks, bicycles, and pedestrians. The purpose of these ambushes is to loot, abduct, restrict traffic movement, and create fear. The tactics used include creating a roadblock, launching an RPG in front of the vehicle, or strafing the vehicle with machine gun fire.[46] After the vehicle is stopped or disabled, the LRA will loot the vehicle and either kill, interrogate, or abduct the passengers. Similarly, raids involve sneaking into an IDP camp, school, or trading centre and then attacking.

The LRA's use of surprise is probably its most traditional guerrilla warfare tactic, almost certainly learned from the former UPDA tacticians. What separates the LRA from many other insurgencies is that it uses these tactics not just against the UPDF but also against civilians, even mostly against them, and thereby creates anxiety among the local population as well as the army. Both raids and ambushes create immediate fear in their targets, especially given the reputation which the LRA has for extreme violence. This is a valuable tactical advantage for the LRA, particularly when dealing with civilian targets. Those attacked are likely to either completely freeze in fear or become very compliant.

---

42 Refugee Law Project, 'Behind the Violence: Causes, Consequences, and the Search for Solutions to the War in Northern Uganda', Working Paper No. 11.

43 Interview with senior government official, Kampala, 10 July 2003 – reported in Refugee Law Project (2004): 20.

44 Joes, *Resisting Rebellion: The History and Politics of Counterinsurgency*, p. 12.

45 Joes, *Resisting Rebellion: The History and Politics of Counterinsurgency*.

46 Recently there has been an increase in the use of landmines in ambushes. (CSOPNU) 'Nowhere to Hide: Humanitarian Protection Threats in Northern Uganda'.

Anxiety

The most exceptional tactic of the LRA has been its ability to maximize unpredictability, which in turn has created a pervasive climate of anxiety throughout the civilian and military population. In particular, the small, dispersed, independent LRA units that roam the sizeable countryside of northern Uganda create a high degree of unpredictability.[47] This results in a situation in which raids and ambushes may literally occur at anytime and in any place across the north. This strategy has expanded in scope over time. Initially, it was possible to track the group of rebels as they infiltrated from Sudan and avoid specific roads or areas for a time by either hiding out in the bush or moving to a more secure trading post or town. Now, all roads beyond a few kilometres of major towns and all IDP camps and trading centres are potentially insecure.[48]

The LRA has also added to its strategic unpredictability by refusing to state or commit to political objectives, or any sort of goal in general. The Refugee Law Project reports, based on extensive interviewing of citizens, government officials, and others in the north, that some respondents variously believed that: the LRA has no political agenda, that there may be an agenda, but that it has not been articulated, as well as consistent confusion as to why Kony claims to be fighting for the Acholi but then kidnapping them.[49] These views were consistent with those encountered during recent fieldwork.[50] Other perceptions included that Kony was trying to take over the government, that he wanted to 'stick it to' the President, that there really is a spiritual reason, and that he is in it for the money. Ex-combatants also showed similar confusion; returnees noted multiple motivations: that the war was to overthrow the government, involved land issues, was punishment for the Acholi people, or was simply about distrust of the government.[51]

Together with this political confusion, the LRA is also notorious for randomly and inexplicably targeting different groups. As one informant put it, 'one day they will say they are going to kill everyone who is riding a bicycle, but then you will ride a bicycle and they will pass you by, but then a week later, they will decide to kill you.

---

47 Confusion and unpredictability are also exacerbated by the fact that the LRA often wear UPDF uniforms. This makes it difficult for civilians and even soldiers to recognize whether someone is a rebel or government soldier at a distance – creating an 'invasion of the body-snatchers effect.' One informant reported that the only way to tell the difference is that the rebels have long hair while the real UPDF have military style haircuts. (Interview with head of the World Food Program officer, Gulu, 13 January 2005.) The problem is compounded by the Sobel, or soldier by day, rebel by night, practices of the UPDF. (De Waal, 'Contemporary Warfare in Africa'.)

48 Reported by Father Carlos Rodriguez. Rodriguez (2003).

49 Refugee Law Project, 'Behind the Violence: Causes, Consequences, and the Search for Solutions to the War in Northern Uganda', Working Paper No. 11.

50 Various interviews in Kampala and Gulu District, January 2005.

51 See for instance Human Rights Watch 'Abducted and Abused: Renewed Conflict in Northern Uganda'.

You can't tell.'[52] For instance, even though the Catholic Church was not a target of attack throughout most of the war, in June 2003, Kony reportedly ordered the LRA to begin targeting it. This order was subsequently followed with multiple attacks on churches.[53] Then, as suddenly as LRA attacks on the Church began, they ended, and now it is again felt that the Church and its employees are relatively safe. Together these tactics have allowed the LRA to construct a climate of fear and anxiety which may potentially affect anyone, anywhere, and at anytime in the north of Uganda.

Comparisons with Insurgency and Terrorism

The LRA has maximized its use of fear along the entire outlined typology. In the ways listed above it has created a pervasive climate of fear which affects everyone in northern Uganda, whether civilian or military. This use of fear by the LRA is different than the ways in which insurgents and terrorists use fear.

Insurgencies rely on fright, but do not otherwise maximize fear in the manner of the LRA. Generally, they use fright as a means to gain strategic advantage. The typical insurgency uses ambushes and surprise raids in order to effectively multiply their forces against the generally much larger forces they encounter in their fight against the state or occupying power. The LRA uses fright in this way as well.

To a lesser extent, insurgencies will also attempt to promote a sense of fear, by trying to create a perception that they are an organization to be feared. However, doing so is impeded by the fact that insurgencies must mobilize public support and this means that the insurgency cannot always exhibit the levels of savagery necessary to create a truly feared organization. While insurgencies may use such fear as a threat, say against civilians who might inform on them, they are more concerned with attracting support than threatening it, for as Walter Lacquer notes: 'no guerrilla movement can possibly survive and expand against an overwhelmingly hostile population.'[54] Since the LRA does not rely on public support, it can afford to alienate and threaten the civilian population.

Insurgents may also attempt to create an environment of anxiety by trying to make it seem that attacks can be made on anyone and come at any time. However, they are impeded in maximizing anxiety by the fact that their goals in the conflict are readily known. The insurgent wishes to overthrow the state, secede from it, or otherwise bring about some specific political end. These groups go through great pains to educate the public, and even the international community, on these goals. The LRA does no such thing; its goals are unspecific and variable. In this way, the LRA can create an atmosphere of anxiety that no true insurgency can possibly reproduce.

Terrorists also do not maximize fear in the same way as the LRA does. A terrorist group does attempt to create an environment of anxiety through attacking civilian

---

52 Interview with local resident, Gulu, 11 January 2005.
53 BBC News, 'Uganda LRA rebel leader "speaks"', (London, 15 April 2004).
54 Walter Laqueur, *Guerrilla: A Historical and Critical Study*, (London, 1977), p. 401.

targets. This creates a high level of anxiety in that civilians feel that they may become targets at any moment. It is such anxiety, combined with the use of spectacle, which defines the terrorist form of warfare. However, terrorists cannot produce the levels of fear and fright that the LRA can. It is the decentralized, shadowy nature of the terrorist organization that works against its ability to produce true fear. The terrorist exists within society and cannot be seen until the moment of the attack. There is no entity to fear in the same way as the definiteness of a psychotic rebel with a machete. Rather, it is the act itself which is feared, and this act tends to last only a split second, with only its after-effects left to deal with.

Similarly, there is no true group, in the sense of an attacking squad of heavily armed teenagers, which can create fright. True, the terrorist attack is always a surprise in the sense that it is not possible to know exactly where and when an attack will occur. But, there is no strategic advantage in this surprise – the terrorist will usually die in the attack or be far away from the seen of the incident. The attack is meant to increase anxiety and create the spectacle by which it promotes its message. It is not intended as a means to momentarily confuse its adversary and thereby to gain an advantage. The exception might be in the more rare instances of multiple, consecutive attacks. However, even when multiple attacks are used, the confusion is not a goal so much as the increase in the impressiveness of the spectacle.

The conclusion is that the LRA is using fear in a way that is very different than that of the insurgent or terrorist. Unlike these groups, it maximizes fear in all ways. But, if it is not using fear to fight an insurgency or for terrorism, what then is it doing? In one sense the LRA's use of fear has given it a strategic advantage. Specifically, its use of fear has acted as a force multiplier against the military. The specific means of force multiplication were noted in the examination above. This has allowed the LRA's small force of 500 to a few thousand men and woman to fight a war against the 50,000- or 60,000-strong government forces of the UPDF and its affiliated militias. However, there is more than just force multiplication behind the LRA's use of fear. If it were only for strategic advantage in battle, the LRA might just be considered an insurgency. Rather the group has taken fear to another level and used it in a way that neither insurgencies nor terrorists can. What the LRA is doing is, in effect, fighting a dirty war.

Dirty Wars

Certain types of war seem to be driven by the logic of creating a general sense of fear within a society. The term *dirty war* is most often used to describe 'campaigns of state-sponsored terror and repression whose goal is to suppress suspected civilian resistance.'[55] As Nordstrom notes, 'dirty wars seek victory, not through military and battlefield strategies but through horror. Civilians, rather than soldiers, are the tactical targets, and fear, brutality, and murder are the foundation on which control

---

55 Carolyn Nordstrom, 'The Backyard Front', in *The Paths to Domination, Resistance, and Terror*, edited by Carolyn Nordstrom and JoAnn Martin, (Berkeley, 1992), p. 261.

is constructed.'[56] In dirty wars, atrocities are purposefully conspicuous and serve the purpose of undermining security, or to use De Waal's phrase, to 'destabilize'.[57] The intention is to create a long-term, anxiety-ridden environment. While the term was typically reserved for state sponsored war, particularly in Latin America, this type of warfare can be practised by non-state actors.

Nordstrom applies the concept of dirty war to a non-state actor, the Resistencia Nacional Mozambicana (RENAMO) in its war in Mozambique.[58] In fighting a dirty war, RENAMO created a culture of fear in which over half a million people died over a decade of fighting. The South African backed RENAMO, a 'rebel movement that has virtually no ideology or popular support',[59] used many of the same tactics in war that the LRA is now using. It was made up mostly of abducted child soldiers and severe atrocities were committed against the local population, including mutilations such as severing body parts. Victims were forced to watch family members being killed and witnesses were left after the atrocity to tell the story. Looting and abductions were also a common occurrence during the conflict.

The description of LRA operations depicted above demonstrates that it is fighting a dirty war – affording it a strategic control over the population of northern Uganda. As Van Acker puts it:

> [T]he LRA must immobilize the population through fear. Terror is a vehicle to project power towards the Ugandan state by creating a state of exception and immobilizing the population, on the fringes of society, effectively enough to enforce a distinction between 'law' and 'unlaw', where rules other than those set by the LRA do not hold. To be effective, terror must be more than a threat which, tragically enough, is confirmed by the daily litany of atrocities. While the desired political change remains non-specific, indiscriminate violence – terror – becomes an end in itself; it generalizes responsibility through the logic of the hostage: since anybody can be hit, anybody can be can be blackmailed by terrorism.[60]

Together the sense of insecurity and anxiety weaken the population's sense of control over their environment and makes them easier to manipulate.

While the LRA does not attempt to control territory in the traditional sense, it does have the ability to effectively control the movements, and thereby the lives, of millions of people who live in northern Uganda. The population of districts in northern Uganda can be essentially 'herded' out of the countryside and into camps at the will of the LRA. They can be forced to provide food or other supplies to the rebels as well as recruits. A profound effect of the LRA's dirty war strategy has been

56 Nordstrom, 'The Backyard Front', p. 261.

57 De Waal, 'Contemporary Warfare in Africa'.

58 Nordstrom, 'The Backyard Front'.

59 Nordstrom, 'The Backyard Front', p. 262.

60 Van Acker, 'Uganda and the Lord's Resistance Army: The New Order No One Ordered', p. 350.

to put political pressure on President Museveni.[61] The immense insecurity present in the north and effective control of the population has made Museveni seem incapable of providing security. Not only does this make him immensely politically unpopular in the north, an electoral issue for him, but it also serves as a major stain on his international reputation, which is otherwise well respected. While the LRA may be committing atrocities on the Acholi population, and is thereby unpopular itself, this does not translate into popularity for the government. In fact, the government is now resented for its inability to provide security.

'An Insane Man's Warfare Strategy'?

Rather than only extolling the brutality of the LRA, this chapter has sought to examine exactly how the LRA has used fear in its conflict. It was found that the LRA's use of fear is not a side effect of an insane man's warfare strategy, it is a systematic goal. The LRA has maximized its threat, effectively used surprise, and created an unpredictable and insecure atmosphere throughout northern Uganda. Moreover, this chapter has illustrated why the LRA has created a climate of fear in northern Uganda. Doing so has given the LRA tactical strategic advantages in its conflict. Beyond these advantages, the LRA has used fear to pursue a strategy of fighting a dirty war. In a sense, fear was found to be both a means and ends in the conflict.

It has also been demonstrated how the LRA differs from both insurgencies and terrorist groups. The LRA uses fear for force multiplication – a rationale in line with its roots as an insurgency. At the same time there is also a terrorist element to the LRA's actions – terror is created and manipulated for political goals. However, neither traditional terrorists nor insurgents take the use of fear to its ultimate end of fighting a dirty war as the LRA has done. In this sense, the LRA's dirty war strategy fundamentally separates it from both insurgencies and terrorist groups – we may even consider it an essentially different form of asymmetric warfare. It has truly gone beyond terror or insurgency. The implication of this is that the analysis of the LRA from the perspective of it being either an insurgency or a terrorist organization is bound to fail.

Indeed, responses to the LRA based on a counter-insurgency framework have failed. For instance, the roots of the Ugandan Government's protected hamlet strategy are in a view that the LRA is an insurgency. A standard form of counter-insurgency is to set up protected villages, such as IDP camps, and the Ugandan Government has facilitated this. However, the strategy has been ineffective. The LRA, unlike traditional insurgents, does not rely on the support and mobilization of

---

61 The climate of fear created by the LRA has wide ranging secondary effects. Fear of abduction has caused many schools to close which has caused a large portion of the present generation of children to go without education. The farming industry has been crushed in a region renowned for its fertility and earlier in the war the cattle population was decimated, in an area where cattle are used as the retainers of wealth. This is all, of course, on top of the severe weakening of the Acholi culture and social structure and the deaths of thousands of people.

the people. Therefore, an attempt to separate the LRA from its (illusionary) public support has had little effect on the efficacy of the organization. Essentially, this is because instead of relying on public support, the LRA relies on public fear.

The counter-insurgency strategy employed by the UPDF has only served to empower the LRA by promoting its ability to incite fear and fight a dirty war. The huge tracts of unpopulated land in northern Uganda have allowed it to operate freely, without the possibility of encountering government informants. This allows the LRA to maintain its ghost-like image and to exaggerate its perceived ubiquity. Since no one sees them coming, it seems like they come out of nowhere and are everywhere – thereby increasing their ability to promote fear.

Similarly, treating the LRA as a terrorist organization would have little real effect on the organization. The LRA is organized and uses fear in a very different way than a terrorist group. It is not a network organization and does not exist within society; nor does the LRA attack in order to create a spectacle. Rather, the LRA exists as a separate, definable group from society which can attack as a whole and continually preys on society. Using counter-terror techniques, such as trying to break up cells would be meaningless and infiltrating the organization would be impossible as its recruitment is based on coercion.

This leads us to the question of how to respond to the LRA. The dirty war which the LRA is fighting is an entirely separate form of asymmetric warfare from terrorism and insurgency. Consequently, it is necessary to develop strategies to effectively combat the LRA, in the same way that counter-terror and counter-insurgency techniques have been developed to respond to these specific types of asymmetric warfare. Once 'counter-dirty war' strategies are developed to respond to the LRA, they can be applied to other types of armed groups fighting a similar kind of war.

The pertinent question to ask is, how can one counter a dirty war? In essence this means asking the question of how one can counter fear. The answer may be found in the usual remedy for phobias. That is, fear must be faced. Making Kony a living, breathing human being in the eyes of people would help in this respect. It is no accident that few outsiders have met him. Bringing Kony into the light would help to end the perception that he is to be feared. This might be accomplished by forcing Kony himself to a negotiating table or even providing simple documentary evidence of his existence, beyond the decade old pictures recycled in newspapers.

Another counter-dirty war approach is to take advantage of the fact that the LRA relies on fear to motivate its troops. The Ugandan Government has done this to a certain extent by instituting an amnesty program, which makes it harder for Kony to convince abductees that they cannot return to society. However, the UPDF needs to go further in taking the fear out of those who want to return. This could be done, for instance, by specifically directing forces to actively aid abductees in escape or by using propaganda techniques to inform those in the bush that they can escape and that the UPDF will help them to do so.

A counter-intuitive approach would be to reverse the protected hamlet strategy. By allowing (or encouraging) the Acholi people to return to the countryside, the true nature of the LRA organization might come to light in that people would see how small an organization it really is. It would also make it a more attractive option for those who are with the LRA by choice to return from the bush, since there would be a normal society to return to, instead of the despair of the IDP camps.

Although it might be assumed that doing this would just lead to a bloodbath, in fact there is little reason to believe that there would be any significant increase in attacks. It is already the case that the LRA can attack the IDP camps at will. Moreover, the LRA is not a very big organization and it seems to be maximizing its attacks already.

This is, of course, only the beginning of a discussion of alternate approaches to confronting the LRA. The ongoing debate between military and peaceful solutions to the conflict in northern Uganda continues to assume that the LRA is either an insurgency or terrorist group. Neither approach seems to be helping to defeat the organization, as the conflict has continued unabated for two decades. By taking a different approach and appreciating that the LRA is systematically creating a climate of fear and fighting a dirty war, it may be possible to make some headway in ending the conflict.

The repercussions of the debate about the LRA go beyond just this conflict. While the use of dirty war as a strategy has been perfected by the LRA, the LRA is neither the first nor the last armed group to rely on it. RENAMO also used very similar tactics as did the Revolutionary United Front (RUF) in Sierra Leone. We can also expect other armed groups now and in the future to use similar approaches, if only because it seems to work. One possible candidate that can be described as fighting a dirty war is the insurgency in Iraq. The armed groups there have also used fear strategically to generate a perception that they are organizations to be feared, they have relied on surprise, and they have created an environment of anxiety. While the comparison is not exact, the very fact that there is no agreement on whether to call the armed groups there terrorists or insurgents does lead toward this conclusion. There may be something to learn from the LRA and its conflict which can be applied to the situation in Iraq as well as possible future conflicts. If only for this reason, the LRA needs to be studied and effective responses found to finally bring an end to its reign of fear.

PART II
Barbarity as Strategy
in Modern Warfare

Chapter 6

Why there is no Barbarization but a Lot of Barbarity in Warfare

Uwe Steinhoff

There is a widespread belief, promulgated in journalistic accounts as well as in some scholarly works, that wars, or at least those wars that are not conducted between states, have become more 'barbaric'. The barbarism of these 'new wars' (or, as they are variously called, 'post-modern wars', 'wars of the Third Kind', or 'People's Wars') allegedly differs from the civility of the 'old wars' in four respects. Old wars, so the story goes:

- were motivated by ideology or collective grievances;
- had broad popular support;
- limited violence to the instrumental; and
- largely respected the distinction between combatants and non-combatants so that many more soldiers than civilians were killed.

The 'new wars', in contrast, are:

- motivated by greed and a desire to loot;
- lack popular support;
- display gratuitous or 'expressive' violence; and
- disregard the distinction between combatants and non-combatants, so that many more civilians than soldiers are killed.

The striking pedagogic neatness of this schematic contrast should at once give one pause. War is such a messy business that these differences between old and new ones seem too clear-cut to be true. However, many theories become popular precisely because they are easy to grasp, not because they are true. The 'new wars' doctrine is one such example.

What is wrong with it? Stathis Kalyvas aptly explains that:

> ... the distinction between 'new' and 'old' civil wars ... is based on an uncritical adoption of categories and labels grounded in a double mischaracterization. On the one hand, information about recent or ongoing wars is typically incomplete and biased; on the other hand, historical research on earlier wars tends to be disregarded. This is compounded

by the fact that the end of the cold war has robbed analysts of the clear categories that had made possible an orderly, if ultimately flawed, coding of civil wars. Accordingly, the distinction drawn between post-cold war conflicts and their predecessors may be attributable more to the demise of readily available conceptual categories than to the existence of profound differences.[1]

Kalyvas's judgement is confirmed by testing the theory against the ethnographic and historical data. We do this briefly with each of the four alleged differences between 'new' and 'old' wars:[2]

1. *Ideology or grievance v greed*: Looting was widespread even in wars that count as prime examples of ideological ones, for example in the Russian and Chinese revolutions. Moreover, the polarization in the Chinese Cultural Revolution developed not so much along ideological lines as along pre-existing cleavages between different clans, which subsequently tended to use ideological vocabulary as a veneer. Micro-studies of the allegedly highly ideological conflicts in Ireland, Vietnam, Peru, and even of the conflict between German occupiers and Russian partisans in the Second World War, show that the choice of sides was more strongly influenced by local (including economic) considerations than by abstract ideological ones.[3] Conversely, conflicts like those in Sierra Leone and Liberia, which are often cited as prime examples of wars motivated by greed and the desire to loot, did in fact involve significant ideological elements and political goals, according to ethnographers Stephen Ellis and Paul Richards.[4] It may also be mentioned that the conflict in Angola started during the cold war and continued after the collapse of the Soviet Union. Before 1989 it was widely considered to be an ideological conflict; afterwards it was generally considered to be a conflict about oil and diamonds. Is it not obvious that here only the perception of the conflict changed, not its nature?

2. *Popular support*: Non-governmental parties in 'new wars' tend to be described as bands of mere thugs with no popular support. The perception of the rebel movement RENAMO is an early example. However, as Kalyvas points out, citing studies by Tom Young and Mark F. Chingono, 'RENAMO enjoyed a considerable level of popular support. This was present in rural areas controlled by RENAMO, where researchers and journalists rarely travelled, rather than in the cities under governmental control.'[5] The same can be said about different factions in the Liberian war, such as NPFL, Ulimo or LDF.[6] Conversely, the extent to which leftist guerrilla

---

1   Stathis N. Kalyvas, '"New" and "Old" Civil Wars: A Valid Distinction?', *World Politics*, 54, October 2001, pp. 99-118, at 99.

2   The discussion of the first three differences is strongly indebted to Kalyvas, '"New" and "Old" Civil Wars'.

3   Ibid., pp. 102-113. See there also for further references.

4   Stephen Ellis, *The Mask of Anarchy: The Destruction of Liberia and the Religious Dimension of an African Civil War*, (London, 2001); Paul Richards, *Fighting for the Rain Forest: War, Youth & Resources in Sierra Leone*, (Oxford and Portsmouth, 2004).

5   Kalyvas, '"New" and "Old" Civil Wars', p. 110.

6   Ellis, *The Mask of Anarchy*.

movements in Latin America or Asia enjoyed widespread popular support has been exaggerated. In particular the Vietcong, often praised for its popular support, relied considerably on coercion of the civilian population.

3. *Controlled and instrumental violence*: It has often been pointed out that the use of child soldiers in the 'new wars' is one reason for their allegedly gratuitous violence. However, great numbers of child soldiers have also fought in such 'ideological' conflicts like the Chinese Cultural Revolution, the Afghan insurgency after the Soviet invasion, the Shining Path insurgency in Peru, and in Guatemala, El Salvador and Nicaragua. Besides, civil wars have always been described as particularly cruel, beginning at least with Thucydides' description of the war in Corcyra. So there is nothing new here.

Western journalists – and the scholars who are satisfied with reading them instead of better-informed ethnographers – are often impressed by the carnivalesque appearance of some fighters in wars like those in Liberia or Sierra Leone. The appearance of fighters dressed up in fancy uniforms or in women's clothes running around with all sorts of talismans is often interpreted as some kind of psychological regression. The same holds true for the widespread use of drugs. Especially with respect to Africa, these interpretations often have a certain culturalist if not downright racist overtone. However, the use of drugs was widespread in old wars and among European soldiers. As Richard Holmes reports, 'nearly half a pint' of brandy per man was served to the infantry divisions of Saint-Hilaire and Vandamme before the task of seizing the Pratzen at Austerlitz. For many years British soldiers got a rum ration shortly before battle.

> At least 10 per cent of Second World War American troops took amphetamines at some time or other, and in 1947 one-quarter of the prisoners in US military jails were 'heavy and chronic users'. American medics often issued dexedrine to soldiers before they went out on night patrol in Vietnam. ... V.G. Kiernan maintained that alcohol was vitally important to the soldiers of colonising powers.[7]

The German First World War veteran Rudolf Binding 'lamented the fact that the cellars at Albert and Moreuil "contained so much wine that the divisions, which ought properly to have marched through them, lay about unfit to fight in the rooms and cellars ... The disorder of the troops at these two places ... must have cost us a good fifty thousand men."'[8] So much for the disciplined, abstinent old European warriors in comparison with the new, primitive, drug-addicted African ones. Regarding their choice of clothing, the Europeans do not necessarily fare better. Here, too, Holmes's findings are extremely valuable:

> On occasion, the escape [from the stresses of war and battle] is compounded by a physical transformation in which the drunken soldier – or even, on occasion, the sober one – dresses

---

7   Richard Holmes, *Acts of War: The Behaviour of Men in Battle*, (London, 2004), pp. 246f.

8   Ibid., p. 250.

up in outrageous garb. On 28 March 1918 Rudolf Binding, sent forward to find out why the advance had slowed up, found 'men dressed up in comic disguise. Men with top hats on their heads. Men staggering. Men who could hardly walk.' Wheeler relates how, after the Battle of Vittoria: 'British soldiers were soon to be seen in French generals' and other officers' uniform covered with stars and military orders, others had attired themselves in female dresses, richly embroidered in gold and silver.' There was a bizarre carnival in the town of Fredericksburg after the battle. Not only was there a good deal of looting and vandalism, but Union soldiers leaped about in women's dresses and underwear.[9]

Finally, whoever thinks that the regression to magic is typical of the 'new wars' rather than of war itself quite simply does not know what he or she is talking about. To quote Holmes one last time:

> In their efforts to load the dice of fortune, and to gain some comfort thereby, soldiers have for centuries cherished talismans or adopted talismanic behaviour. During the First World War, wrote Paul Fussell, 'no front-line officer or soldier was without his amulet, and every tunic pocket became a reliquary. Lucky coins, buttons, dried flowers, hair cuttings.'[10]

It seems, then, that all these aforementioned forms of behaviour that are taken as signs of barbarity if displayed by non-European fighters are too characteristic of wars generally, including old European ones, to constitute something new.

In addition, the gratuitousness of some forms of violence to be witnessed in the 'new' wars – but not only in them, by the way – especially of violence directly targeted at civilians, is more imagined than real. As Kalyvas points out:

> The massacres in Algeria were often highly selective and strategic, as was the violence used by RENAMO. Young found that the most extreme atrocities were part of a carefully drawn—and largely successful plan—to battle harden young, mostly forcibly conscripted young guerrillas. Likewise, atrocities committed against the population at large were concentrated in southern Mozambique, where the FRELIMO government had a strong base.[11]

Similar observations have been made about Sierra Leone.[12]

4. *The combatant/non-combatant distinction*: It is true that in the wars fought in Europe during the nineteenth and earlier twentieth centuries between 70 and 80 per cent of those killed were soldiers. It is not true, however, that the reversal of the ratio between killed combatants and killed non-combatants that we witness in the so-called new wars in the Balkans or in Africa is indeed something new. It is, actually, mirrored by the casualty ratios of the imperialist wars of the nineteenth centuries, which were on the one side fought by proper European states or by the USA. And it was the armed forces of these states that were responsible for the reversal, not their 'primitive' enemies. In the Second Philippines War of 1899-1902, 4,000 US

9   Ibid., pp. 252f.
10   Ibid., p. 238.
11   Kalyvas, '"New" and "Old" Civil Wars', pp. 115f.
12   Richards, *Fighting for the Rain Forest*, p. xx.

troops, 20,000 insurgents and 200,000 civilians were killed. This is almost a 10 to 1 ratio of civilian to military deaths.[13] The Spanish war of occupation in Latin America, for example, cost the lives of 35,000 Spanish soldiers and reduced the population of Venezuela, New Granada, Ecuador and Mexico by 770,000 people.[14] Most of the dead, of course, were civilians. The colonial wars of Britain, France and Germany did not put too much weight on the distinction between combatants and non-combatants, either.

Thus, the thesis that a 'new' kind of war has appeared that differs – not least in terms of 'barbarity' – from the wars of the nineteenth century or from the allegedly more ideologically oriented civil wars that were fought during the cold war is wrong. There is simply no evidence for some kind of deep historical shift to more barbaric 'new wars'.

However, it might be worthwhile to point out that a specific form of barbarization is discernible in some European and especially in the American armed forces of today as compared with those of the eighteenth and nineteenth centuries and of the two world wars. This barbarization has not so much to do with an increased readiness to kill civilians – this readiness was already there before, at least in wars outside Europe – but with an extreme unwillingness to risk the lives of one's own soldiers (which was not there before). Far from being civilized, this unwillingness was shared with the proverbial barbarians, the Mongol hordes of Ghengis Khan, who owed their victories – contrary to the conventional wisdom – not so much to superior courage or primitive 'fierceness' as to superior technology and tactics. As John Keegan reports:

> ... in their management of animals they showed a matter-of-factness – in mustering, droving, culling, slaughter for food – that taught direct lessons about how masses of people on foot, even inferior cavalryman, could be harried, outflanked, cornered and eventually killed without risk.[15]

This preference for killing without risk and for the 'eschewal of heroic display'[16] was at the time taken by the Christian Occident as the mark of the barbarian. This barbarism, however, is now manifested by the way Western powers prefer to conduct wars. The preferred method is to fire missiles from far away and to bomb the enemy from a great height, remaining untouchable oneself. This way of fighting makes it even more difficult to distinguish between civilians and combatants. It increases, therefore, the extent of so-called 'collateral damage', that is, of civilian casualties. But then it is quite hypocritical if the same powers that engage in such practices vilify terrorists as barbarians.

---

13 Errol A. Henderson and J. David Singer, '"New Wars" and Rumours of "New Wars"', *International Interactions*, 28, 2002, pp. 165-190, at 175.

14 Jörg Friedrich, *Das Gesetz ds Krieges: Das deutsche Heer in Rußland 1941-1945. Der Prozeß gegen das Oberkommando der Wehrmacht*, (München, 2003), p. 68.

15 John Keegan, *A History of Warfare*, (London, 2004), p. 213.

16 Ibid.

To this observation a member of the UN High-level Panel on Threats, Challenges and Change recently retorted that every army in the world puts the lives of its own soldiers before the lives of foreign civilians. However, it is, first, unclear how this tribalism squares with the universalist liberal values the USA, Great Britain and other European nations purport to defend. Second, it might also be true that every army, insurgency or militant group puts its own cause before the lives of the civilians on the other side. But if the USA and European nations are justified in 'collaterally' killing scores of foreign civilians in order to keep their own soldiers safe, why are sub-national groups not justified in killing civilians in order to advance their causes? Not even the infamous Doctrine of Double Effect – which is wrong anyway – is of any help here, for it does not defend the disproportionate collateral killing of civilians. Yet, given that the life of a US or European soldier is not worth more than the life of an Iraqi or Afghan civilian, the bombing campaigns of the Allies or the 'coalition of the willing' are disproportionate. In short, one does not have to cut off other people's hands with cutlasses or to degrade them by forcing them to feign homosexual acts; a good old bombing campaign might be quite sufficient. As Paul Richards nicely puts it:

> There is little if any analytical value, it seems to me, in distinguishing between cheap war based on killing with knives and cutlasses, and expensive wars in which civilians are maimed or destroyed with sophisticated laser-guided weapons. All war is terrible. It makes no sense to call one kind of war 'barbaric' when all that is meant is that it is cheap.[17]

The Psychology of Barbarity

The fact that there is no barbarization of warfare as such has the important consequence that it makes little sense to try to explain such a (non-existent) historical shift. Rather, it would make much more sense to explain why some wars are more barbarous than others. This, however, requires some understanding of the basic psychological dynamics that makes war such a fertile ground for barbarity and inhumanity in the first place. How barbaric a war is depends on how well these basic psychological dynamics are held at bay or are even manipulated. In the remainder of this chapter I use the findings of the Milgram experiment, the Stanford prison Experiment and the so-called Terror Management Theory to shed some light on these dynamics.

In the 1961 *Milgram experiment* participants were recruited via a newspaper advertisement to take part in a study of memory. The experiment required them to give electrical shocks to a 'learner' (who was actually an actor) whenever this learner made a mistake. At the beginning of the experiment the participant received a sample 45-volt electric shock and was then told that the voltage of the shocks to be administered to the learner were to be raised by 15 volts after each mistake. Each time the participant wanted to halt the experiment, the experimenter prodded him

---

17 Richards, *Fighting for the Rain Forest*, p. xx.

on with remarks such as: 'The experiment requires that you continue. Please go on.' 'At 300 volt the actor bangs on the wall separating him from the subject. At 315 volt the actor bangs on the wall again and after that there is no further response from him and he stops giving answers to the memory questions.'[18] A meta-analysis of repeated performances of the experiment undertaken in the USA and elsewhere found that the percentage of participants who inflicted voltages up to the maximum of 450 volts lay between 61 and 66 per cent, regardless of time and place. Compliance decreased with the degree of immediacy of the victim and increased with the degree of immediacy of the authority. In some experiments the participant was joined by one or two additional 'teachers' (who were actually actors). When these teachers refused to comply, only four of 40 participants still complied. If the additional teachers complied, only three of 40 participants did not. Women were as obedient as men, although they seemed to experience higher levels of stress.

In the 1971 *Stanford prison experiment* participants were also recruited by newspaper ads. Of the 70 participants, 24, who were deemed to be the most psychologically stable, were selected. They were predominantly white, middle-class young males. The group was at random divided into 'prisoners' and 'guards'. The 'prison' was a prepared basement at Stanford University. Guards were given wooden batons, military-style uniforms and mirrorshade sunglasses. Unlike the prisoners they could return home during off-hours, although some preferred to 'work' even during off-hours. Prisoners had to wear ill-fitting muslin smocks without underwear, tight-fitting nylon pantyhose caps, and small chains around their ankles, and they were assigned numbers. The only formal guideline given was that no physical violence was permitted. The experiment got quickly out of hand and had to be stopped after only six days of the planned two weeks. The guards tormented the prisoners, denied some of them food as a form of punishment and even imposed physical punishments, forced nudity and homosexual acts of humiliation. Several guards became more and more sadistic and most of the guards were upset when the experiment was cut off. Of 50 outsiders who had seen the prison, only one objected to the conditions.[19] Incidentally, the study was funded by the US Navy to explain conflict in its and the Marine Corps' Prison systems, so the US forces should have heard about this study and drawn some conclusions from it before Abu Ghraib could happen.

Terror Management Theory posits, in the words of its founders:

> that human beings' highly sophisticated cognitive capabilities render us aware of the ... inevitability of death. This awareness engenders the potential for overwhelming terror – terror that is assuaged through the construction and maintenance of culture: humanly constructed beliefs about the nature of reality that confer symbolic or literal immortality by providing a sense that one is a person of value in a world of meaning...

---

18 'Milgram experiment', http://en.wikipedia.org/wiki/Milgram_experiment, accessed on 26/1/2005.

19 'Stanford prison experiment', http://en.wikipedia.org/wiki/Stanford_prison_exper iment, accessed on 26/1/2005.

Empirical support for TMT has been obtained in more than 150 studies conducted by independent researchers in at least nine countries. ... Research has ... established that subtle reminders of death produce increased clinging to and defense of one's cultural world-view. These tendencies include the following: greater affection for similar others or those who uphold cherished cultural values and greater hostility toward different others or those who violate cherished cultural values, heightened discomfort when handling cherished cultural icons in a disrespectful fashion, sitting closer to a person who shares one's culture and farther away from a foreigner, and increased physical aggression toward someone critical of one's cherished beliefs. Additional studies revealed some important individual differences that moderate the effect of mortality salience on worldview defense. Specifically, people who have high self-esteem (dispositional or situational), who hold liberal world-views that stress tolerance (or for whom the value of tolerance has been primed [that is, brought to their situational awareness]), or who are securely attached are less prone to disparage those who are different from themselves following mortality salience induction.

Subsequent research ... led us to make a distinction between the proximal and distal defenses that occur, respectively, in response to conscious and unconscious concerns about death. When mortality is made salient, direct rational *proximal* psychological defenses are activated to reduce conscious awareness of death, by instrumental responses ... such as distracting oneself from the problem, denying vulnerability to the threat, or emphasising the temporal remoteness of the problem. Once the problem of death is out of focal attention but still highly accessible, terror management concerns are addressed by *distal* defenses, by bolstering faith in the worldview (e.g., by derogation of those who violate or challenge one's worldview and enhanced regard for those who validate the worldview) or by enhanced self-esteem striving (e.g., behaving altruistically if being helpful is an important aspect of one's self-concept).[20]

This is, in fact, a long passage packed with information, and some of its findings or implications should be spelled out. I wish to do this with reference to the Abu Ghraib scandal. As Pyszczynski, Solomon and Greenberg emphasize, 'people need concrete manifestations of their values and beliefs to serve as a reminder of the power, protection, and security that these more abstract psychological entities provide.' The attacks of 9/11 therefore constituted a double blow: by killing thousands of Americans in one strike they made Americans aware of their mortality and, by simultaneously striking prominent American symbols like the World Trade Center and the Pentagon, they undermined the confidence in their world-view that was needed to fend off the disconcerting psychological effects of the mortality salience. Therefore, their world-view had to be defended with a vengeance, as it were, which explains some of the extreme reactions to dissenters at the national and international levels. The guards in Abu Ghraib will hardly have been free of these effects. The fact that Abu Ghraib happened some years after 9/11 is of little relevance if, as is safe to assume, the constant reminder of 9/11 played some part in their mobilization and motivation by superiors. Moreover, they did not even need the reminder of 9/11. The insurgency

---

20 Tom Pyszczynski, Sheldon Solomon and Jeff Greenberg, *In the Wake of 9/11: The Psychology of Terror*, (Washington, 2003), p. 190f.

in Iraq, the ongoing killing of American soldiers or of members of private military companies around them made their mortality salient enough. However, American prison guards in Iraq are not as directly and constantly confronted with their own mortality as combat soldiers. Soldiers who have to patrol in a hostile area or who are even involved in combat will quite often be *consciously* aware of their own mortality – and hence react, according to TMT, with proximal responses like the one described in the passage quoted above. American prison guards in Iraq, on the other hand, will more often be *unconsciously* concerned about their deaths. And such an unconscious concern about one's own mortality breeds the *distal* responses described above, among them the derogation of those who violate or challenge one's world-view – as Muslim or even Muslim insurgent detainees do – and an enhanced striving for self-esteem. Given that many of the photographs taken in Abu Ghraib and elsewhere depicting the humiliation of Iraqi prisoners had the character of trophies, it is clear that this humiliation was a way to enhance one's own self-esteem: 'see how powerful I am and what I can do with these ungrateful, Islamic Iraqi bastards.'

The lessons to be drawn from the Milgram and Stanford prison experiments are also quite clear. The Stanford prison experiment shows how easily power can corrupt, especially power over people who are – albeit only symbolically – de-individualized or dehumanized. The most salient message of the Stanford prison experiment could be summed up in the slogan: 'It's the barrel, stupid.' The barrel and not a few bad apples. The bad barrel produces bad apples, and not only a few but loads of them. Given the fact that the US Navy funded the Stanford prison experiment, there are two possibilities. One is that those responsible for setting up the prisons in Iraq – this, of course, includes the most high-ranking officials in the US armed forces – did not know about the experiment. Such ignorance and incompetence would, however, constitute itself as uttermost negligence and recklessness. The other possibility, which seems to me the more likely one, is that they did know quite a lot about it. In that case, the conditions that led to the torture and the degrading and humiliating treatment in Abu Ghraib and elsewhere were actually desired and condoned at the highest levels in the American army and in the American administration.

The Milgram experiment, finally, shows how easily normal people can be brought by even the mildest pressure from the authorities to do the most outrageous things. In fact, there need not be any real pressure at all, for in anticipatory obedience many people try to execute what they *think* the authority wants from them. In other words, the authority does not have to explicitly say 'Torture the prisoner' in order to get the prisoner tortured. There are other ways to convey one's wishes.

Fortunately, the Milgram experiment, the Stanford prison experiment and Terror Management Theory not only go some way in explaining how events like those in Abu Ghraib can happen, but also give very useful clues on how to prevent them. For one thing, neither the Stanford nor the Milgram experiment shows that guards in prisons will necessarily humiliate or even torture prisoners. In most prisons in the US, for example, the brutality stems not so much from the guards but from the prisoners themselves. In fact, the guards are pretty much under control.

Consider the Milgram experiment. It shows how far obedience to authorities can go. This obedience, however, can be used not only to make people torture or humiliate others but also to make them abstain from such activities. Imagine that the experimenter in the Milgram experiment had said to the participants: 'By the way, because of another experiment to be conducted when you go, the learner will be connected right now to this electric shock machine on your desk. So please, keep your hands off that machine in order not to inadvertently give the learner a shock.' I am not aware that this variant of the Milgram experiment has been undertaken anywhere, but I think it is safe to assume that, apart from the odd psychopath, no participant would give the learner any electroshock under these conditions. Analogously, a simple clear order: 'Do not torture or humiliate the prisoners!' would have done a lot to prevent the abuses in Abu Ghraib. But it seems that no such order was given. Also, the situation in the Stanford prison experiment deteriorated because the experimenters did not intervene. That must, under the prevailing conditions, have had the effect of implicitly condoning the abusive practices. The situation would have been very different if the experimenters had issued the general guideline, 'Treat the prisoners decently' and called guards off in response to abusive practices.

TMT, finally, shows that pre-screening of candidates for a job as prison guards should look out for people who have high self-esteem, hold liberal world-views that stress tolerance and are securely attached. However, some doubts are in order as to whether these are really the character traits that the armed forces, even in liberal-democratic countries, are looking for. But perhaps they should look for them. At the very least, the value of tolerance should be primed, that is, brought to the conscious and situational awareness of the staff in military prisons. Empirical studies show that this has a mitigating effect on the tendency, brought about by mortality salience, to derogate those who dissent from one's own worldview. This, however, works only with people who attach any importance to the value of tolerance. This should be the case, to a greater or lesser degree, with most Americans and citizens of other liberal-democratic states. This priming of the value of tolerance could be effected in the American context, for example, by simply putting up placards with texts like, for instance: 'Tolerance and human dignity are the highest values in the American constitution. As Americans, we respect even the dignity of our enemies or prisoners, even if our enemies might not.' Also, they could put up passages in which the founding fathers stress the value of tolerance and emphasize human dignity. To my knowledge, nothing of this sort has been done in Abu Ghraib. Maybe the paper was too expensive.

Ignorance vs Political Will

To answer the two questions in the title of this essay: there is no barbarization of warfare in our time because the phenomena that we witness now and that some of us take as new and particularly barbarian are neither new nor in all cases necessarily particularly barbarian. Nevertheless, there is a lot of barbarity in warfare, for two reasons. First, power corrupts, and a lot of human beings, as the Stanford prison experiment shows, have a certain tendency to become cruel if given the chance under certain conditions. If they are not only given the chance but are prodded to abuse others by an authority, very few resist. The second reason is that many authorities – mostly for instrumental reasons – do prod subordinates to commit atrocities and abuse prisoners, or fail to take measures to prevent atrocities and abuse or, for that matter, disproportionate 'collateral damage'. Of course, even a well-meaning belligerent authority will not be able to prevent all abuses. There is little to be done about the odd psychopath even in peace, let alone in war. But it can very significantly reduce the likelihood of the occurrence of atrocities and abuses, often, as the psychological findings show, by quite simple means. If these means are not employed, or not to a nearly sufficient degree, outright stupidity and ignorance might be an explanation. Lack of political will is another.

Chapter 7

Barbarity and Strategy

Graham Long

It is self-evident both that barbarity is common in war, and that it is a bad thing. The abuse and mistreatment of combatants, prisoners and civilians has always been part of the history of war. The continued presence of barbarity, despite our repugnance, is something that requires explanation. I aim here to contribute to such an exploration through an examination of the strongest groundings for justifying barbarity – asking whether there is ever a moral case for committing barbarous acts.

This chapter asks whether barbarity can ever be considered as a form of strategy or policy in warfare and, if so, whether it can ever be morally justified. I want to suggest that we should not reserve the term barbarity for senseless or random horror – which I take to be obviously morally wrong – but also apply it to horror of a purposive and discriminating kind. If barbarity can be a means to an end then the possibility of a defence of barbarity based on the validity of those ends must be considered. It becomes an open question whether barbarism can be justified as an exception to the rules of war. A further, wider, justification of barbarity as policy can be founded in diverse understandings of what *counts* as barbarity. This defence argues that different cultures will have different accounts of barbarism, and so judging between them will be difficult. Before I examine these questions, however, some ground needs to be cleared. Is it possible to talk of strategic barbarity? And isn't all war barbaric?

First, I should say something about the nature of barbarity. Historically, barbarity has been opposed to civilization; it is what the civilized find inferior. Barbarity, as I use it here, retains some of this sense, but we should resist easy identifications of the 'civilized' and 'barbarians'. The ancient Greeks applied the term to any non-Greek, and Plato, for example, argues that barbarians deserve worse treatment in warfare than his civilized fellow Greeks: the nature of barbarians suited them only for slavery. I propose to oppose barbarity not to civilization but to *humanity*. Barbarity is, on this account, behaviour that is both inhumane and inhuman. This definition has two components; it is action that (a) shocks our consciences, and (b) is fundamentally morally wrong.[1] The

* This chapter was presented at the Barbarisation of War Conference, University of Wolverhampton. I also gratefully acknowledge the financial support of a British Academy Postdoctoral Fellowship during the writing of this chapter.

1 It would perhaps be better here to say that barbarity involves the violation of human rights. Nevertheless, I choose this formulation to avoid clashing with accounts of war that regard combatants as suspending or losing their right to life – e.g. Michael Walzer, *Just and*

first component of this definition captures our sense of repugnance at barbarity. The second means that what must be involved is a real and deep violation of morality – not merely something that we find offensive. Adopting this kind of definition will have consequences later in the chapter, because of its reference to shocking *our* consciences. When, precisely, should everyone's conscience be shocked? This reference is intentional, and constitutes an ambiguity with the limitation of barbarity in warfare.

On this definition, war is barbarous in its very nature. I will say slightly more about this later in the chapter, but for now, we can note that war involves a commitment to the killing, intentional and unintentional, of an indeterminate number of fellow human beings. It harms both the victims and perpetrators of violence. If this does not shock our consciences, it should. If we do not consider this cruel, monstrous or inhuman, we should.[2] War is indeed hell, as Sherman wrote, but it is not so by accident. Because war is barbarous in its very nature, the only consistent way to end all barbarity in war is pacifism.[3] So, how can we talk about war becoming more barbarous, or of particular acts as atrocities? I take it that when we want to talk meaningfully about the impermissibility of barbarous acts in war, we must mean acts that stand out from the ordinary horror of war, that are in some way 'beyond the pale'. Picking out the particularly barbarous acts in something that is inherently cruel is a second problem that will recur in my chapter.

Thinkers of the just war tradition have done precisely that, setting out a set of rules that limit the conduct of warfare.[4] Such rules require three things. The first is proportionality in the use of force – the idea that destruction must be in proportion to the military advantage gained. The second is non-combatant immunity from attack – civilians, because they are not threats, cannot be targeted. The third is a restriction on horrific means of waging war, such as chemical and biological warfare. These rules, I want to suggest here, broadly track the distinction between the ordinary

---

Unjust Wars, (New York, 1977), pp. 138-145. There are, to my mind, problems with such accounts. Most importantly, we seem to lose the sense that there is something wrong with ending another human's life or inflicting massive pain and injury. Whilst combatants may lose their right to life, it seems difficult for them to declare themselves free of a duty not to kill other humans.

2   I exclude other kinds of violence here. Duelling, for example, strikes me as potentially non-barbaric, and so wars that are settled through this mechanism (for example, those described in Irish mythology) are outside my analysis, as perhaps are conflicts fought using entirely non-lethal means.

3   Nevertheless, there are reasons why pacificism – the complete rejection of war – is not an attractive option given the world as it is. I take it that most of us can imagine cases where what was at stake would be worth fighting for. On an individual level, defence of our family might be one such case: at an international level, humanitarian intervention to prevent genocide might be another. This is not to say we should not be *pacifistic*, that is, go to war only reluctantly, and work to eliminate it from our world.

4   E.g. Walzer, *Just and Unjust Wars*; James Turner Johnson, *Can Modern War be Just?*, (Westford, Mass., 1984).

horror of war and excessive barbarity. They demarcate what shocks our conscience the most. They do not, however, turn a war into a fair fight.

It is important to note that these moral rules of war may not be the same as the restrictions imposed by international law. Throughout this essay, I make my argument largely in moral terms. International law, to my mind, is best perceived as a highly imperfect reflection of moral principles, skewed by the interests and shared history of different countries, most notably the great powers. To say something is morally required is generally to say that there may be a good case for this being reflected in law. However, we can make moral judgements independent of the law, and indeed judge the law to be morally flawed.

So, the first distinction necessary is between the everyday horror of war and those actions we deem especially barbarous or inhuman, a distinction tracked by aspects of the rules of warfare. Having said something about the nature of barbarity, and the way that it is reflected in the moral rules of war, I begin the chapter proper by making a distinction amongst acts of barbarity. This second distinction is between two *kinds* of acts of barbarity.

The easiest way to think of barbarity is to conceive of it as something unplanned, something done in the heat of the moment – the random and immediate response of the combatant to a situation. Something about barbarity, perhaps left over from its opposition to civilization, suggests unthinking, unreasoned violence. This account might seem an attractive one, for it limits the scale of the crime. As an example of this, military investigations found the prisoner abuse scandal at Abu Ghraib in the second Gulf war to be confined to a few soldiers who disobeyed orders. Allegations that torture was permitted or encouraged higher up the chain of command persist, however.[5] Even in cases of 'rogue combatants', there will often be systemic elements of policy present, so it is wrong to focus purely on the act as somehow isolated or random. The training a combatant receives, their experience of war, and the way they conceive of their enemy will all be determinants of their response in the field.

Cases of barbarity as unthinking cruelty – the malfunctioning of an individual's moral compass – can be contrasted with strategic or purposive acts of barbarity that reflect the orders of the regime or of military planners. The distinction might be thought of as that between barbarity that occurs in war despite orders, and that which occurs precisely because of orders. Barbarity is not just something that can occur between individual combatants in the field, but also something that can inform a regime's entire mindset or strategic thinking. Thus, the German campaign on the Eastern Front in the Second World War reflected orders from the highest level to

---

5   The report blamed 'misconduct ... by a small group of morally corrupt soldiers and civilians', compounded by a lack of leadership and discipline: see George Fay, Anthony Jones, *Investigation of Intelligence Activities at Abu Ghraib*, (2004), http://www.defenselink. mil/news/Aug2004/d20040825fay.pdf. By contrast, some argue that responsibility for the abuse extends to the top of the administration. Ambiguities exist over the extent to which the US administration permits, or encourages, means verging on torture (see, for example, www. crimesofwar.org).

exterminate Slavic populations, and to place no value on the life of Slavs or Jews whether military or civilian.[6] Such a policy can be described as barbaric in the same way as the acts of individual soldiers who carried it out.

The first class of barbaric actions are in some ways less troubling: whilst horrific, there is at least a ready response. They lie outside the rules of war, and, in being against orders or military codes, are clearly subject to sanctions (at least in principle). That is, the conventions governing warfare already provide ways to respond to these atrocities, though imperfectly. In fact, this is one reason why ascribing responsibility for acts of barbarity to a few individuals is an attractive prospect. It allows 'closure'; the offenders can be tried and found guilty. Strategies of more pervasive barbarity, however, pose a more difficult question, because these tend to involve at least a limited rejection of rules of war. They challenge the idea that the rules of war should apply, or else hold that they should be broken.

I want to insist that there are many cases where we can confidently and clearly condemn acts as barbaric, where there is no question over how deeply they shock our consciences and how obviously and without justification they violate basic human rights. I take this to be a basic cornerstone of our experience of barbarity, and it is not something I wish to deny. The most obvious of these cases are where barbarity is found in the end or point of the war. Some political goals, such as genocide, are barbaric. In the rest of this chapter, however, I am concerned with some hard cases and difficulties for the condemnation of barbarity. I pick out three potential problems. First, the chapter looks at how we can decide the line of excessive barbarity that the rules of war should track. Second, I consider the objection that barbarity can be used in emergencies as a 'last resort'. Third, I consider the possibility that accounts of barbarity can legitimately differ from society to society.

Excessive Barbarity?

If all warfare is barbaric, we might think it difficult to form a notion of excessive barbarity that should be prohibited and punished. An approach starting from the premise that 'war is hell' has not always been opposed to barbarism if it speeds up the resolution of the war in our favour. This is an important objection to my approach. However, rather than showing a problem with my approach alone, a survey of the kinds of barbarity found in war will instead show more general difficulties in arriving at a clear line of prohibition.

I want to begin by considering two responses to this problem suggested by the work of Thomas Nagel. In his influential paper *War and Massacre*, Nagel offers an argument against the conclusion that anything goes in war. Nagel offers what he terms an 'absolutist' account of restrictions of warfare, one that forbids us from doing certain things to people. His absolute restrictions mean certain actions are ruled out altogether. On his account, we are morally obliged to show respect to others even

---

6   Antony Beevor, *Stalingrad*, (London, 1999), pp. 54-61.

when fighting them. The first response suggested by Nagel's work is that a kind of warfare characterized by respect would not be inherently barbaric. Absolutism requires treating other combatants with respect as persons, even in combat. We need to be able to offer others justification for our actions. Nagel writes, 'One could even say, as one bayonets an enemy soldier, "it's either you or me".'[7] I think Nagel is correct to say that justification is significant, and this is a theme I will return to. The important thing here, however, is that this approach seems to offer the possibility of a non-barbaric warfare; one where we respect others as persons even as we kill them in their role as combatants. On this approach, there is a clear division between ordinary war, and the barbarity that occurs where we fail to respect others in this way.

However, this appearance is mistaken. The fact we can offer some kind of justification for war, or for our action in war, does not render war non-barbaric in nature. Having a justification or 'just cause' for war, such as self-defence, does not make the actions one performs less barbarous in themselves. Instead, the moral wrong done is outweighed by the moral good of the cause – it makes the course of action less barbarous, *all things considered*. War can only ever be the lesser of available evils. Killing someone is a violation of their most basic interest; though we might argue that their right to life is suspended by virtue of them having 'combatant' status, the content of the act, of taking another human life, is always horrific. But it is not just the humanity of those killed or maimed that is violated by war. The lives of those who kill, even when in self-defence or by accident, can be ruined by the trauma and guilt that follows.

A second consideration concerns the scale of war. Whilst individual soldiers may construe their fight as one of self-defence, where they can justify their killing to their opponents, what is barbaric in war is in part the *scale* of the killing. War removes individuality; in Nagel's terms, following Kant, people become treated as means to a strategic or political end. Even though individual soldiers might respect other individual soldiers, all are in the end following orders. They kill out of necessity only because they are put into a situation where they have to. For these reasons, I do not think that Nagel's account offers the possibility of a war where combatants emerge without dirty hands. Combatants' experiences of the horror of war seem to bear this out. This is not to deny that there are better and worse ways of waging war, just to deny that we can have a genuinely respectful war.

The second argument we can take from Nagel holds that, though we might allow that ordinary war is not civilized, there is still a difference in kind between ordinary and barbarous warfare. An example Nagel uses illustrates this point. Nagel argues that starvation, poison, flamethrowers, napalm and the like are 'particularly cruel weapons' 'designed to maim or disfigure or torture … rather than merely stop.' These weapons, for Nagel, 'abandon any attempt to discriminate in their effects between the combatant and the human being.'[8] I want to maintain that such weapons are

---

7 Thomas Nagel, 'War and Massacre', in C. Beitz, J. Cohen, T. Scanlon (eds), *International Ethics*, (Princeton, 1990), pp. 53-74, p. 67.

8 Nagel, 'War and Massacre', p. 71.

different only in degree from the 'ordinary' weapons of war that attack 'combatants' rather than 'humans'. They violate the same basic human interest in life, but do so in a more conscience-shocking way. Any lethal weapon of war does not merely end the threat posed by the combatant, but the life of the human being also. Conventional weapons cause terrible injury in the process of 'stopping' enemy combatants. Long-range area attacks (artillery barrages, for example) inflict horrific injuries without any of the immediate or direct qualities Nagel is looking for in war. Nor are cruel weapons a feature purely of modern war. There are many candidates for methods of warfare that might be cruel enough to warrant prohibition. For these reasons, there is no sharp line between weapons that attack men and are thus impermissible, and weapons that attack soldiers, which are thereby rendered ordinary or non-barbaric. Instead, the line we have to draw is based on difficult conscientious reflection.

We might also add that the *intent* behind the weapon is as important as the weapon itself. Even the most basic weapon can be used to maim or disfigure as well as kill. The intent behind the attack – whether to defeat by killing every enemy, or instead accept the surrender of some, for example – is where we can distinguish between combatants and human beings. Whilst two different attacks may kill equal numbers of enemy combatants, the intent manifest in the attacks will also have a role in determining how horrific we find them. Furthermore, the rules of war (as they have traditionally been formulated) have no objection to mass killing where necessary – there might be nothing especially wrong with the *scale* of killing. As Hurka notes, principles of just warfare do not require us to incur casualties equal to our enemy; instead, we are morally permitted to kill whatever amount of enemy soldiers necessary to save the life of one of ours.[9]

It might be thought that the situation is clearer with regard to rules establishing the immunity of civilians from harm: civilians are simply not allowed to be the direct targets of attack. However, orthodox thinking on the rules of war allows for civilian casualties as foreseen consequences of an attack where the intended target is a military one. That is, provided the force used and casualties caused are proportional to the military advantage gained, war can legitimately cause civilian deaths. This is not the same as civilian deaths being the only or primary aim of an attack. Nevertheless, on this account, a bomber pilot can legitimately drop a bomb knowing that it will end civilian lives. Nagel is uneasy about this distinction between intending to kill, and killing as a foreseen consequence, and I believe he is right to be.[10]

Thinking about war is partly concerned with consequences, and the consequences of these actions are importantly the same – the death of friendly and enemy

---

9   Thomas Hurka, 'Proportionality in the Morality of War', *Philosophy and Public Affairs*, 33, 2005, pp. 34-66.

10   Nagel, 'War and Massacre', p. 61. This distinction is often referred to as the principle of double effect. It should be noted that not all philosophers of war affirm the same version of the principle. Walzer, for example, adds strict qualifications: Walzer, *Just and Unjust Wars*, pp. 151-159. Thompson rejects it completely: Judith Jarvis Thompson, 'Self-Defense', *Philosophy and Public Affairs*, 20, 1991, pp. 283-310.

combatants or non-combatants. To determine what counts as excessive barbarity we must consult and consider our moral intuitions that tell us some ways of killing or certain intentions are worse than others, and apply these in balance against military necessity. We determine the lines of prohibition by working through our moral convictions about war. The rules that result are not comfortable, settled principles grounded in a fundamental difference in kind between barbaric and non-barbaric warfare, but instead anxious compromises with the logic of war. One consequence of this understanding of the rules of war as the weighing of arguments is that justified barbarity becomes possible. In considering whether particularly barbarous acts can be justified or not, we are not doing anything different from our ordinary thinking about war. A second consequence is grounded in the crucial role our conscience plays in these judgements. This is important because, as the diverse world we live in demonstrates, people can come to different conscientious verdicts. The following two sections address branches of these concerns. The first considers the view that morality may permit barbarity if the intent or end is good enough. I will term this a *consequentialist* defence of barbarity, since it argues that the good consequences may outweigh the bad, and indeed outweigh the horror of the act itself. The second argues that different cultures or consciences will draw different lines with regard to barbarity – this is a *relativist* defence of barbarity.

The Consequentialist Defence: Barbarity as a Means to an End

Can barbarity as a violation of the rules of war be countenanced on the ground of necessity? The appeal of using barbarous actions is twofold. First, the rules of war that express the prohibition on barbarity can make victory more distant; they impose costs on combatants. One appeal of suspending or violating these rules is that these costs are thereby avoided. Barbarity can thus be the 'low-cost' option, and as we shall see, there are times when this is especially attractive. The second appeal comes from the effect of barbarous acts on the enemy's will to fight through 'shocking the conscience' of the opposition. Barbarous acts have an effect on the morale of the enemy. This can be seen, for example, in the justification offered for British bombing of German cities in the Second World War. Barbarity also demonstrates the will or conviction of the combatants.[11] For these reasons, barbarity may appeal as a strategy. Committing those acts we consider barbarous – sometimes precisely because they will be considered barbarous by our enemies – can seem a strategically sound proposition.

This may be the strategic attraction of barbarism, but for the consequentialist defence to be persuasive, the end has to be worth taking such steps. Justifying barbarous policies by reference to a barbarous end will not work. Thus, these kinds of qualified defences of barbarism work best when something important is at stake – when we conceive of a war as a defence of our most basic moral values. The

---

11 As Kurtz says in *Apocalypse Now*, recounting an act of barbarity on the part of the Vietcong, 'I thought my god, the genius of that. The genius. The will to do that. Perfect, genuine, complete, crystalline, pure. And then I realised they were stronger than we.'

Second World War is often invoked as this kind of case. The defence of our political community, or of human rights, or simply avoiding the triumph of great evil, are perhaps such values.

Again, Nagel's account is useful here as a starting point. Despite Nagel's absolutism forming a barrier to ever committing acts of barbarism, his account – perhaps paradoxically – still leaves room for 'ends-means' reasoning. Where Nagel's absolute prohibitions clash with considerations of the best consequences, we are left in a 'moral blind alley': 'the world can present us with situations in which there is no honourable or moral course for a man to take, no course free of guilt and responsibility for evil.'[12] In particular, Nagel admits that in situations of deadly conflict, the moral dilemma can be 'acute'. So, rather than offer a prohibition against any possible use of unsavoury means in warfare, Nagel instead is offering only a particular account of how we should understand the moral dilemma that results. It is an account which says that considerations of consequences, and considerations of respect for persons, cannot be combined in a single calculus. It does not, however, insist that respect for persons should always win out. In that regard, it leaves room for permissible barbarity.

There are two kinds of relevant necessity here. The first kind of necessity is more tactical; it considers how a set of military objectives can be achieved. In the case of Abu Ghraib, for example, let us assume that what took place was not unthinking abuse but instead interrogative torture, at least in part. In the context of an ongoing resistance movement striking from within a civilian population, a shortage of information was crippling the effort to put down the insurgency. The argument for a policy of torture would be that this violation of the rules of war would save the lives of Coalition and Iraqi personnel and civilians in a situation where other options were very limited or costly. This kind of defence of barbarity is best put in terms of a calculation. That is, it weighs the value placed on the life of friendly combatants and civilians versus the prohibition on barbarous acts, such as torture. Such justifications will be difficult, since rules of war are designed precisely as breakwaters against this pressure of military necessity. The logic, though, can appear inexorable (though not in the Abu Ghraib case). If a war can be sped up, a just cause served, and hundreds of thousands of lives (on both sides) saved through an attack that violates the rules of war, it is hard for military leaders, and their political masters, to dismiss such an option out of hand.

The second, larger kind of necessity finds expression in Michael Walzer's idea of 'supreme emergency'. Where a country is threatened with 'extinction or enslavement', necessity becomes the overriding consideration and 'necessity knows no rules'. For Walzer, the British bombing campaign against Germany in the Second World War is a paradigm case of supreme emergency. Even though it violated the rules of war by being insufficiently discriminate in its targeting, it was justified, when it was the only option for prosecuting the war, precisely because it was the only option and the costs of defeat were so great.[13]

---

12 Nagel, 'War and Massacre', p. 73.

13 Walzer, *Just and Unjust Wars*, pp. 255-268.

Such justifications of barbarity face problems, however. First, it would seem that almost any political leader could construct a case that avoiding their country's (or cause's) destruction means contemplating barbarism. Second, Alex Bellamy notes in a recent article that invocations of barbarous means as a last resort always rest on a 'historical fallacy' – that is, they occur in complex situations with many available options which become reduced to an 'either-or' choice. He writes, 'the fallacy, I argue, is the claim that political and military leaders confront one of two possibilities in supreme emergencies: refuse to break the rules of war and face destruction; or directly target enemy non-combatants.'[14] Usually, argues Bellamy, there are specific reasons why the choice gets narrowed down to leave only the options of barbarity or else our own destruction. These reasons are often bad ones – for example the preference of vocal advisors – and so the choice is unduly narrowed, missing out other options which involve less barbarity. It can be argued that this combination of factors – of the potential for the abuse of supreme emergency, and the unduly narrow way in which leaders tend to construe the choice – should lead us to reject a supreme emergency defence of barbarism.

However, I want to maintain that at least some barbarity *could* be justified on the ground of supreme emergency. The first reason is that we can imagine cases where there are genuine moral dilemmas, as Nagel suggests. In such circumstances, the impending evil may be sufficiently great to override concerns about barbarous means.[15] This can be true even when choices are unduly narrowed; though leaders should be open to all the options, there is nothing here to suggest that the barbarous option could never be chosen even in such non-ideal conditions. Bellamy's argument applies best to the subclass of barbaric actions that involve the intentional mass targeting of non-combatants. Here, he makes a convincing argument that such targeting seldom, if ever, has strategic value; it is not the most *effective* last resort. We should note, though, that in such situations Bellamy himself countenances a move towards barbarity – he believes that in supreme emergencies we can weaken the rules covering foreseen civilian deaths, so that the lives of enemy civilians will weigh less than our soldiers.[16] This effectively weakens our duty to preserve civilian life, and permits us to launch much more devastating though inaccurate attacks against the enemy.

It is always possible to construct counter-cases where barbarity is the only or the most effective last resort – where the proper calculation of losses, risks and benefits favours violating the rules of war. Consider the well-thumbed case of torturing a suspect to reveal where a devastating bomb is hidden. There is, of course, a very good chance that the information the suspect would reveal under torture would be false. Nevertheless, if it was our only realistic chance of finding the bomb, an

14 Alex Bellamy, 'Supreme Emergencies and the Protection of Non-Combatants in War', *International Affairs*, 80: 5, 2004, pp. 829-850, p. 840.

15 I am sympathetic to the view that defences of supreme emergency work best when what is at stake is something bigger than a concern for mere state survival, as I discuss shortly.

16 Bellamy, 'Supreme Emergencies and the Protection of Non-Combatants in War', p. 848.

argument for it as the lesser of available evils – though under very tight restrictions – is endorsed by many moral philosophers.[17] Just as it is possible for barbarity to sometimes be the right choice, though, it is possible also for barbarity to be resorted to too quickly or easily, for the wrong reasons, or in the defence of the indefensible. Thus, to say that barbarity is sometimes legitimated by the perception of a supreme emergency is not to say that it is always justified.

The perception and judgement of military and political leaders is crucial here. Leaders will be left in the kind of moral dilemma that Nagel describes, weighing their responsibilities to their country and to humanity more generally, the harm that will be done and the chances of success. How we regard disagreement about these decisions is crucial. John Rawls' account of 'reasonable disagreement' can, I believe, give us an appropriate model. Rawls notes that factors such as differences in the interpretation of the same evidence, or attaching different weight to available pieces of evidence, allows for the possibility of reasonable disagreement – that is, one where two parties can arrive at two different verdicts, neither of which are clearly incorrect.[18] Both are reasonable, but neither can be shown to be the right one as far as the other is concerned. I want to maintain that these enormously difficult questions of when to invoke supreme emergency – how great and immediate a threat, what the consequences will be of action or inaction – are subject to these kinds of factors, and thus we can expect reasonable but different responses to them. The scope for disagreement is compounded because the decision-maker must take into account all future potential losses arising from the action. Lastly, the more moral worth attached to the end, the more barbarity can be considered permissible in the means.[19] But the moral worth of the ends themselves will also be subject to reasonable disagreement. For these reasons, all we can expect from military and political leaders in such dilemmas is a good-faith, conscientious and well-reasoned decision. Given the immense moral and factual difficulties in weighing all potential dangers and benefits, we should expect persistent and hard-to-resolve disagreement.

Let me sum up my argument here: it has two components. The first is that if we want to retain the right to invoke barbarous actions or means in conditions of 'supreme emergency', then we cannot consistently hold that barbarity is always the morally worst option. Instead, there are cases – though they may be very limited in number – where we must be open to considering barbarity the lesser of two evils.

The second adds uncertainty into this account. If all we can expect of people deciding upon the options is a good-faith conscientious decision that is not clearly irrational, then we have to expect people to employ barbarous means in situations where we would not. That is, there will be situations where the use of barbarity will seem to us unjustified, but to others it will not. On my account, the conditions

---

17 E.g. Henry Shue, 'Torture', *Philosophy and Public Affairs*, 7:2, 1978, pp. 125-143.

18 John Rawls, *Political Liberalism*, (New York, 1993), pp. 54-58.

19 Thus, the defeat of a threat to all political communities or of the rights of everyone might serve to justify more violations of the rules of war than a threat purely to our national sovereignty.

under which barbarity can be justified are worryingly subjective. We will have situations where two people can regard the same calculation in very different ways. As I examine in the next section, I think this should matter to our condemnation of barbarity and how we regard the moral guilt of those who carry it out.

The problem of potentially justified barbarity has its roots, it seems to me, in the singular nature of war as barbarity. Since war is just mass killing, distinctions between how we kill, and whose lives to take versus whose we sacrifice are, in a way, unavoidably fluid. We should expect the possibility of trade-offs and violations of these boundaries – the very nature and pressures of war tempt us in this direction.[20]

The Relativist Defence: Barbarity as Culturally Determined

At the beginning of the chapter I invoked the idea of barbarity shocking our consciences. Because it is *our* conscience being shocked, the second objection I consider argues that there is room for relativism here – relativism being the idea that different societies (or individuals) will have different conceptions of precisely what it barbarous, and that no single one of these conceptions can be shown to be the correct one.[21]

This second objection charges that acceptable barbarism will vary between moral frameworks, and so there is no universally accepted standard of barbarity. In effect, this objection argues that any moral rule we might offer which prohibits what we consider barbaric is either grounded in cultural bias, and thus ethnocentric, or else broad enough to be subject to vagueness and indeterminacy.[22] This relativist worry amplifies our concern about where to draw the line between excessive barbarity and the legitimate conduct of warfare. From within another world-view, the lines to be drawn and judgements to be made may look very different.

Diversity in how societies regard killing and war cannot be denied. As one example of this, consider the principle of Samurai ethics – one that found echoes in the Japanese treatment of Allied captives in the Second World War – that surrender is dishonourable and so prisoners should not enjoy the respect of their captors. Consistent with this, dishonoured Samurai would expect to take their own lives if they failed in their duty, and this rule would also govern them if captured.[23] Such an attitude to prisoners approaches the rules of war from a very different angle from the idea, which forms one basis of the contemporary humanitarian law of war, that

---

20 As Shaw has argued, war carries in it the seeds of degeneration into something even nastier, Martin Shaw, *War and Genocide*, (Oxford 2003).

21 The problem that war is essentially barbaric also affects our response to culturally specific accounts of barbarity. There is no clear difference in kind that all cultures can agree upon.

22 These same kinds of criticisms apply to inter-cultural moral norms, notably human rights, much more generally. I do not have the space here to discuss this larger debate or the specific ways in which such charges can be answered.

23 The code of *bushido* and its application is a complex subject, and my characterization of this principle here is only a very rough one.

prisoners (once they lay down their arms) are morally immune to further violence. It is fair to say that the Samurai ethic attached a very different value to life and death, and in particular dying the right kind of death, compared to our own culture.

As a second example, some classic Islamic traditions of rules of war lay down different rules from what we might term the 'western' tradition of just war. On some interpretations of the Koran and other sources, adult males of opposing non-Islamic regimes, even if they surrender, can be put to death.[24] Furthermore, the western tradition of just war itself contains diversity. McMahan's recent work, for example, argues that moral guilt for the war is as morally significant as the threat posed in determining who is a legitimate target.[25] By weakening or removing the prohibition on targeting non-combatants, this offers a very different account of the rules of war.

It does not automatically follow from the *fact* that there are these diverse approaches that we must re-evaluate our account of why and when barbarity is wrong. We need not be left with a kind of nihilism that says 'anything goes'. Instead, our response should depend on the interpretation that we offer of these differences. A crude relativism would argue that all these different approaches can justifiably lay claim to be the 'true' approach to just war ethics. Therefore, *none* of these approaches can really be the only correct one. I believe we should reject such a crude relativism, because it leaves us with no response to moral horror. On this account, the Nazi could insist that we just agree to differ about whether genocide is wrong, and so I take it that such an account would be inadequate. The approach we choose must instead make it possible for us to criticize and condemn others.

However, we should accept that it will be difficult to conclusively justify our account of barbarity to others – that is, to furnish others with reasons that they would be irrational to reject. We should not always be troubled when our condemnation is rejected – perhaps it is rejected on faulty grounds, or by inconsistent and incoherent world-views. Nevertheless, there may remain cases where the best reasons we offer for rules of war will seem unintelligible, or else poor, in the light of others' way of looking at the world. I do not want here to lay out all possible characterizations of such situations, or provide a detailed account of what these standards of justification and proof might consist in. The important point is that this kind of situation matters; in our everyday lives, and in liberal politics, the idea of providing a justification for our actions is an important one. Roughly speaking, we should be hesitant about imposing our view on others within diverse political communities where we cannot justify it to them. The difficulty in justifying or *proving* a particular account of barbarity is something that we should be sensitive towards.

---

24 See, for example, the discussions in John Kelsay, 'Islam and the Distinction Between Combatants and Non-combatants', in J.T. Johnson, J. Kelsay (eds), *Cross, Crescent and Sword*, (Westport, 1990), pp. 197-220, e.g. p. 203; Sohail Hashmi, 'Interpreting Islamic Ethics of War and Peace', in T. Nardin (ed.), *The Ethics of War and Peace*, (Chichester, 1996), pp. 146-168, pp. 162-164.

25 Jeff McMahan, 'The Ethics of Killing in War', *Ethics*, 114: 4, 2004, pp. 693-733.

When we meet radical difference on questions of barbarity this should cause us, first, to reflect upon our own account of the rules of war, and to reaffirm why we think it is correct. It also highlights the importance of at least attempting to provide justifications for our verdict; it will not always be enough to offer condemnation alone, but instead a deeper engagement – repugnant though it may be – with the competing justification offered. Lastly, the degree to which someone's act of barbarity is carried out conscientiously – rather than through wilful viciousness or psychopathy, for example – should affect how much blame we attach to the actor. However, this sensitivity is not the only thing determining our response to barbarity in war. We still possess our moral convictions about when barbarity goes 'beyond the pale', and if, on reflection, we remain confident in them then they should be applied. On this account, our judgements of others' practices in war are made more complex, but are not rendered worthless, when those others possess a different culture.

A Constant Emphasis on Justification

In this chapter, I have argued that strategic barbarity needs to be taken especially seriously because it exists beyond, rather than within, the presence of the rules of war. It questions whether those rules can justifiably be applied, or whether they should be suspended, overridden or indeed ignored because they are not 'our' rules. I have canvassed two problems with the easy condemnation of barbarity. Neither of these objections force us to abandon our moral outrage at barbarity in war, or our efforts to limit it. However, I have also suggested – troublingly – that there are limited and particular circumstances in which barbarity cannot clearly be shown to be unjustified. In particular, others will make different but reasonable judgements on supreme emergency conditions, and have different but reasonable accounts of what counts as barbarity – ones that we disagree with but cannot clearly show to be wrong. These, I have suggested, should cause us to pause for reflection, if only slightly, before unequivocal condemnation.

My analysis has implications for barbarity in contemporary conflicts, and here I will highlight three. We can begin with the distinction between strategic and arbitrary barbarity as a useful analytical tool. Purposeless barbarity is by definition without justification. However, treating barbarity like this stops us from asking the difficult questions. I have investigated one such set of questions, concerning potential justifications for barbarity as policy, here. Another important line of enquiry, which I have not discussed for reasons of space, concerns the extent to which our capacity for barbarism in warfare is the result of training or wider societal influences. Purposive barbarity reflects the nature of war. On my account, the morality of a war itself rests on a calculation over whether the goals of a war can justify the barbarity it will necessarily involve. More localized proportionality calculations, of whether specific actions can be considered because of their contribution to victory, reflect this fundamental one.

Second, I have explained why we should expect persistent and potentially irresolvable disagreement over when barbarity can be justified. The ongoing 'war on terror' waged by the USA and its allies will be especially prone to this kind of disagreement. The nature of a conflict against terrorism, where success is defined by the prevention of devastating attacks against civilian population centres, lends itself with worrying ease to the kind of means-ends reasoning and proportionality calculations I have described. Advocates of the use of torture or detention without trial, for example, will argue that these practices are justifiable precisely on the ground of supreme emergency. Political leaders must weigh these excesses versus the potential risks. Furthermore, because terrorists do not wear uniforms, and live and hide amongst non-combatants, a war against terrorists will continue to involve causing harm to civilians, both by accident and as foreseen consequence.[26] My approach allows for the possibility that such measures might, on balance, be justifiable – our discomfort notwithstanding. Thus, our response to these instances of barbarity cannot always be easy condemnation.

Third, because war is barbaric in nature, we should resist the temptation to focus on particular instances of excessive barbarity at the expense of indifference to war generally. It might appear that war is more sanitized than it ever has been. Casualties on the side of coalition countries in recent military operations in Kosovo, and even Iraq, have been relatively light. However, the media has an important role in shaping our perceptions of warfare. By and large, the Western media toned down or censored images of the fighting in Iraq, showing rubble, but few bodies.[27] We should resist the idea that modern warfare, apart from occasional outrages, is somehow clean.

On the account I have offered here, there will be cases where we cannot offer others a conclusive justification for prohibiting particular cases of barbarism. This does not mean that we should cease to condemn barbarity. Instead, our condemnation should be accompanied by a constant emphasis on justification. This means winning round others to our account of the limits of war, consistency in the application of our standards, and an engagement in the detail of potential conflicting justifications. Thus, our understanding of the rules of war – and the norms that international society imposes to reflect them – should reflect not only our moral outrage at acts of barbarism but also the difficulty sometimes present in justifying this outrage. My account of the rules of war as unsteady and difficult, and their violation as complicated by considerations of diversity and disagreement, is not an optimistic one. This is appropriate: a resort to war should never be the home of moral comfort or easy certainties.

---

26 My point here should be taken only as a general one. The 'war on terror' is a contested idea, and it should not be identified easily or straightforwardly with the second war against Iraq. Ironically, one pressure serving to make war more barbaric in contemporary conflict is the perception that public opinion will not tolerate substantial friendly casualties, since there can be a legitimate fear that casualties will mean not just a setback, but military action being curtailed (Hurka, 'Proportionality in the Morality of War'). Concern over soldiers' deaths may serve to shift the balance of risk to civilians.

27 David Campbell, 'Representing Contemporary War', *Ethics & International Affairs*, 17: 2, 2004, pp. 99-108.

PART III
The Barbarity of Contemporary Culture

Chapter 8

Taking the Gloves Off and the Illusion of Victory: How not to Conduct a Counter-Insurgency

David Whetham

Plato stated in about 375 BC that the point of war is not just to win, but to make a better peace.[1] This is a sentiment that has been echoed by many, from Augustine in the early fifth century to Michael Walzer in the twentieth, but the logic of this simple statement is still lost on many. While the essential nature of war always remains the same, it can manifest itself in many different ways. The current Global War on Terror is just one such manifestation. It is (or at least we are told it is) some kind of global insurgency against Western values, power and way of life. There are many who would claim that democracies have made the job of defending themselves too difficult. Why should we be bound by rules or norms that our opponent shuns? Are the principles of proportionality and discrimination and the laws that flow from these simply moral luxuries that the West can no longer afford if it wishes to defend itself? In this chapter, I will try and demonstrate that the idea of 'taking the gloves off' in response to anything from a small scale local insurgency up to and including the threat of global terrorism, is politically ill-advised, militarily counter-productive and, in simple terms, just plain bad strategy.

George Patton held that 'there is only one unchanging principle in warfare: that is, to inflict the greatest amount of death and destruction upon the enemy in the least time possible.' This is a sentiment that can be seen expressed in US strategy from the days of Sherman and Grant and can still be seen in contemporary US policy in Iraq such as the devastating assault on Fallujah. The moral objection to this attitude can be summed up in the simple question: surely, if you have the choice between defeating your opponent with many casualties or defeating them with few, is it better to kill fewer people? A strategy based on always employing the maximum amount of force will rarely achieve this.[2] Proportionality and discrimination are the two key concepts that can be found in the Just War Tradition's view on the correct conduct of

---

1 Plato, *Laws* I.

2 This also holds true of the idea of radical force protection where any potential risk to friendly forces is reduced even if this is at the expense of increased civilian casualties. It may

hostilities. Proportionality refers to only using an amount of force that is required to achieve a military objective and no more, while discrimination refers to the idea that only those who are directly engaged in hostilities may be deliberately targeted and that there is a moral duty to avoid non-combatant injuries or deaths where possible. The moral case for adhering to these ideas has been made effectively many times.[3] However, the reason that the concepts represented in the Just War Tradition have such an enduring quality is not based solely on the strength of its moral or ethical arguments, but on the fact that they also represent sound strategic advice.

Matching the Means to the Ends

A successful military strategy must accept that a military victory and achieving a satisfactory end state are not the same thing at all, although for a military victory to actually mean anything, the two must be closely connected. There is a clear distinction between conflict termination (e.g. standing on the flight deck of the USS Abraham Lincoln and proclaiming major combat operations are over) and conflict resolution (e.g. a stable and democratic Iraq that is no longer seen to pose a threat). Sometimes a military victory can come at the expense of a political settlement and this makes any military victory hollow at best, disastrous at worst. To provide an historical example, if George Washington had slaughtered the British forces at Yorktown instead of capturing them, it is probable that British public opinion would have been so outraged that every available resource would have been committed to defeating the rebel colonists. Whatever the eventual outcome of that hypothetical confrontation may have been, it is clear that the conflict could potentially have gone on for many more years in spite of, or in fact because of, what would have been a military success. Instead, Cornwallis and the men under his command were granted an honourable surrender and the rest, as they say, is history. To give a more contemporary example, the methods employed by the Russians in Chechnya at the moment also bear this idea out. The excessive and often apparently arbitrary use of indiscriminate force by the Russian military has ensured that no military operation can deliver a peace that would be acceptable to either side.[4] No matter how successful the Russians are in any military sense, victory will remain illusive.

British counter-insurgency (COIN) doctrine, built on a huge amount of experience in 'small wars', contends that this type of attritional response will be counter-productive. Rather than simply trying to liquidize as many insurgents as possible through the application of massive force, the British approach has been to prefer

---

mean fewer 'friendly' casualties in the short run, but this chapter will argue that such tactics can often be counter-productive.

3   For the best contemporary treatment of this subject, see Michael Walzer, *Just and Unjust Wars – A Moral Argument With Historical Illustrations* (New York, 2000).

4   There are many accounts that demonstrate this, but few more succinctly than Steve Crawshaw, 'Military Activities and Human Rights', in Patrick Mileham (ed.) *Whitehall Paper 61: War and Morality* (London, 2004), pp. 131f.

a slightly 'softer' approach and the application of the minimum amount of force required to do the job. Partly this has been down to the strong UK liberal tradition, and partly this has simply been a recognition that the UK army is traditionally fairly small so what force there is available has to be used wisely: the carrot is better than the stick approach. The six general UK COIN principles are:

1. Political Primacy and Political Aim – the recognition that there can never be a purely military response to an insurgency.
2. Coordinated Government Machinery – political, economic, psychological and military tools must work together.
3. Intelligence and Information – this is how the above tools are targeted.
4. Separating the Insurgent from his Support – either physical or moral separation from support base.
5. Neutralizing the Insurgent – using only as much force as is strictly necessary.
6. Longer Term Post-Insurgency Planning – a recognition that there are no quick fixes to insurgencies.[5]

This 'enlightened' British response to insurgency has been learnt the hard way and there are many examples of where excessive force has been employed in the past and has been counter-productive on the political level. I will limit myself to just two examples: Aden and Northern Ireland.

British forces in Aden in the 1960s were attempting to defeat a growing insurgency against imperial rule while at the same time trying to manage a graceful withdrawal from Empire. Operations against insurgents in the Radfan hills around Aden had some level of military success: the insurgents were cleared for some time from the area. However, politically, the operations were a complete disaster because of the methods employed. These included proscription in which after a warning had been given to the people who lived in an area to leave, anybody left in that area was considered a hostile target. This of course targeted insurgent and innocent men, women and children, their livelihoods and way of life, indiscriminately. It did clear the area of insurgents, but when the military operation was over, the area was simply reoccupied as there had been no attempt to conduct any kind of recognizable hearts and minds campaign to actually win over the local population. Collective punishment was also used and this further alienated the people. Rather than defeating the insurgency, unsurprisingly, support actually grew and the insurgents were simply displaced, ironically to the very place that the British were trying to defend – Crater City. Clearly, any short lived military success did not contribute to the achievement of the overall political aim. It is difficult to see how the word success can even be used in this context.[6]

---

5   'Counter Insurgency Operations (Strategic and Operational Guidelines)', *Army Field Manual* (July 2001), Vol. 1, p. B-3-2.

6   In this campaign, the political fallout could not even be blamed on the press for they were largely barred from the area and were generally fairly deferential in the way that they did

In Northern Ireland, a large portion of the Catholic population initially welcomed the arrival of British troops who had been sent there to protect the civilian population from violence. Tactics such as predominantly searching Catholic homes for guns and so on undermined the impartiality of the military. Next came the disastrous policy of internment without trial (it is difficult not to draw parallels with the anger caused in the Muslim world by Guantanamo Bay). In the early hours of 9 August 1971 the British armed forces and Royal Ulster Constabulary (the RUC) rounded up 342 suspected Republicans. Although loyalists had been responsible for much of the trouble, none were detained. Internment was widely perceived as an attack on the whole Northern Catholic community.[7] Few would look back and see this as anything other than a gross mistake as the British had undermined their legitimacy and the population was starting to turn on them. The events of Bloody Sunday in 1972, when security forces opened fire on a civil rights march killing 13 people, went on to further undermine their legitimacy in a way that was almost impossible to come back from. It was clear that overt military means were never going to be successful.

British COIN doctrine has been shaped by such experiences and now clearly recognizes that there will never be a purely military solution to an insurgency. Defeating any insurgency requires a mixture of political, economic, psychological and military means. These factors, taken together, act in the 'struggle for men's minds' rather than towards achieving a straightforward military objective.[8] This was a lesson that had also been learnt by US troops long before the disasters of Vietnam even though it appears to be forgotten each time the US goes to war and has to be relearned again and again. In particular, the US Marines have traditionally been the only branch of the US military to look at the implications of fighting 'small wars'. Their 1940s Small Wars Manual has some fascinating advice that still appears strikingly relevant today. In particular, the importance and primacy of the political situation was clearly understood. For example:

> Small war situations are usually a phase of, or an operation, taking place concurrently with, diplomatic effort. The political authorities do not relinquish active participation in the negotiations and they ordinarily continue to exert considerable influence on the military campaign.[9]

It was made very clear that the application of purely military measures may not be sufficient to restore peace without addressing the fundamental economic, political or social causes of the conflict.[10] It was also to be made very clear that the US was

---

report events. For background to this period, see J. Paget, *Last Post: Aden, 1964-67* (London, 1969) and C. Mitchell, *Having Been a Soldier* (London, 1969).

7   http://news.bbc.co.uk/hi/english/static/northern_ireland/understanding/events/internment.stm accessed 21 July 2005.

8   See Frank Kitsen, *Bunch of Five* (London, 1977), in particular, pp. 281-298.

9   *Small Wars Manual: United States Marine Corps* (Washington, Government Printing Office, 1940), p. 1-7.

10   *Small Wars Manual*, p. 1-9.

not at war with the host country: 'the goal is to gain decisive results with the least application of force and the consequent minimum loss of life.'[11] The purpose of the US military was 'friendly', to restore normal government, establish peace, order and security and, when accomplished, 'leave the country with the lasting friendship and respect of the population.'[12] These and many other sections would be very at home in contemporary UK COIN doctrine, even though the original US manual was written in 1940! Clearly these principles are not new: Gallup polls were being used as far back as the period following World War II in occupied Germany to gauge public reaction and tailor policies to suit where appropriate. This meant the occupying forces on the ground had a very clear idea about what the people were thinking and could thus pre-empt problems that could foster dissent, undermine legitimacy and lead to insurgency in the first place.

This type of successful approach recognizes that insurgencies are defeated not necessarily by killing all of the insurgents, but by persuading enough of the ordinary people that they do not wish to support the insurgents. This denies the insurgents legitimacy while increasing your own, thus depriving the insurgents the ability to be able to move among the people as naturally as fish in water, as Mao Tse Tung so eloquently put it. Staying within the rule of law is essential during any military operation, but arguably, it is even more important during a COIN campaign where maintaining legitimacy and winning hearts and minds are so important.[13] The UK *Manual of Armed Conflict* makes clear that 'necessity cannot be used to justify actions prohibited by law. The means to achieve military victory are not unlimited.'[14] This is not some extra element 'bolted on' to a military strategy as an afterthought, but is also a recognition that successful military strategy must recognize and incorporate these ideas implicitly. The very first chapter of the UK Army Field Manual concerned with COIN doctrine is titled 'Aspects of the Law', stating clearly and unambiguously that 'In whatever capacity troops are employed they must always operate within the law.' It is noted with some concern that the contrasting section in the current US manual is tucked away in Appendix J – the very last section before the index. Some of the sentiments expressed show that the counter productive nature of 'taking off the gloves' is understood in theory:

> Since the very goal of counterinsurgency operations is to help maintain law and order, those conducting counterinsurgency operations must know and respect the legal parameters within which they operate. Those who conduct counterinsurgency operations

---

11 *Small Wars Manual*, p. 1-17.

12 *Small Wars Manual*, p. 1-17. The Manual notes that as well as winning over the local population, the occupying force had to consider domestic US opinion and be prepared for anti-interventionist propaganda from the press or even Congress. It was also important to take into account the view of other countries as negative reactions from foreign powers could impair trade relations or other vital interests of the US.

13 Again, this was certainly not lost on Mao who developed strict rules, such as paying for food taken, etc., for ensuring that the population were won over rather than alienated.

14 UK MOD, *The Manual of the Law of Armed Conflict* (Oxford, 2004), p. 23.

while intentionally or negligently breaking the law defeat their own purpose and lose the confidence and respect of the community in which they operate.[15]

However, one cannot help feeling that, intentionally or not, because of where it has been placed, this is almost an afterthought rather than a recognition of the centrality of these ideas in any successful strategy. This is particularly true when seen in the light of another quote from the US manual: 'All counterinsurgency operations comply with law of war principles *to the extent practicable and feasible.*'[16] Clearly, when fighting an implacable opponent who will be immune to any type of persuasion or negotiation and who feels that it is a religious duty to kill or die in the attempt, there is very little that can be done in response other than to detain or, most likely, kill them before they kill you. However, this still has to be done within the rule of law or legitimacy will be undermined or lost completely. Once legitimacy is challenged in this way or through other use of excessive means, public support will drain away and the insurgency triumphs.

Much has been written on the errors made in US strategy in Vietnam but I think this extract really highlights what this chapter is about:

> May 25, 1969 'Top Secret ... memo from John Paul Vann, counter insurgency ... 900 houses in Chau Doe province were destroyed by American air strikes without evidence of a single enemy being killed ... the destruction of this hamlet by friendly American firepower is an event that will always be remembered and never forgiven by the surviving population.[17]

When proportionality and discrimination and the rules that flow from these time-honoured concepts are abandoned, the military operation is very likely to become counter-productive. In this case the report is talking about directly alienating support of the people that the US was supposedly there to protect.[18] It is easy to see how a situation like this can develop out of an approach such as the Pentagon's Body Count strategy employed in Vietnam. The logic here stated that as long as there were more of them dead than us, the strategy must be working. This type of perverse logic led to the now infamous US Army report filed after the shelling of Ben Tre in South Vietnam: 'It was necessary to destroy the town to save it.'[19]

---

15 HQ, Department of the Army, *FMI 3-07.22 Counterinsurgency Operations*, Oct 2004 expires Oct 2006, J-2.

16 Ibid., J-4. Emphasis added.

17 For a full account, see John Pilger, *Heroes* (London, 1986).

18 General Lewis W. Walt, who commanded the First Marine Division in Vietnam told reporters in 1970 that it had taken them a long time to realize that 'you just can't go in and wipe out guerrillas', adding that the Marines eventually discovered that they had to win the people over. See Ronald Schaffer, 'The 1940 Small Wars Manual and the Lessons of History', *Military Affairs,* April 1972, pp. 46–51.

19 *New York Post*, 7 Feb 1968.

Public Opinion as a Centre of Gravity

The US experience from Vietnam also brings out another aspect about 'taking the gloves off': that it risks losing the support of the people for whom you are waging the conflict.[20] It is a given in contemporary military operations that your principle centre of gravity is likely to be domestic public opinion. This is exactly as it should be in a functioning democracy and whilst there will always be dissenters to any kind of military activity, it is also true that no democratic government can survive committing its armed forces to fight in a cause that is no longer accepted as just or necessary, or right, or in the national interest, etc. by an overwhelming majority of the population of that country. The photo of the burned little nine-year-old girl running away from her village that had been bombed with napalm by South Vietnamese forces (8 June 1972) is one of the iconic images of the twentieth century. It earned the photographer, Nick Ut, a Pulitzer Prize, and probably did more than any other single thing to bring home to the American people what was being done in their name.[21] Opposition to the war in Vietnam was already widespread in the US, but the effect of this picture was enormous. The effect of public opinion in a democracy cannot be underestimated. In an age of global media, this is even truer. The 'strategic corporal' effect is the idea that even the actions of an individual right at the bottom of the chain of command can have a political strategic effect, undermining support at home (e.g. photos of prisoner abuse in British bases in Iraq – a demonstration that sound doctrine does not guarantee adherence to it) or in the wider global community (e.g. internment indefinitely without trial in Guantanamo Bay). Allowing the gloves to be taken off anywhere in the chain of command is not only bad strategy, but it can also undermine the reasons for which you went to war in the first place. The support of the people in whose name military activity is carried out is essential in a functioning democracy. We saw a recent example of this in the 2004 elections in Spain when the PSOE Socialists were elected on a platform of withdrawing from Iraq following opinion polls (before the Madrid bombings) that showed that over 90 per cent of the Spanish people opposed the invasion.[22] Beyond the domestic arena, international public opinion cannot be ignored either. It can have an enormous impact on the stability of coalition forces, trade, diplomacy, etc. A country risks making itself an international pariah at an enormous cost.

Events in Fallujah between 2003 and the end of 2004 provide another example of this Ground Hog Day-type problem with learning lessons from previous mistakes. The huge assault on the city by US forces to quell a high profile insurgency was an effort to remove a challenge to US enforced peace and stability and to prevent the insurgency spreading across the whole of Iraq. Interestingly, few people have looked

---

20 This is discussed very effectively in Michael Walzer, *Arguing About War* (New Haven, 2004).

21 The little girl in the picture, Kim Phuc, is now a peace activist and UN Goodwill Ambassador, living in Canada.

22 http://www.cnn.com/2003/WORLD/europe/03/29/sprj.irq.spain/ accessed 21 July 2005.

back to how this situation developed in the first place. Some of the seeds for this confrontation were sown in an earlier encounter in April 2003 in which 13 protestors were killed and as many as 75 were wounded by soldiers of the 82nd Airborne. Those who died had been demonstrating against the use of a school by the US troops so that pupils could return. Some were upset that night vision goggles were being used by the soldiers as there was a rumour that they were being used to spy into where people were sleeping.[23] While it obviously makes a difference at the tactical level, on the political level, just as with Bloody Sunday, it is pretty much irrelevant if there were shots fired from the crowd. The simple fact is that a number of civilians died in what appeared to be an unprovoked and totally disproportionate response by military forces, caught on film and broadcast around the Muslim world (although hardly reported in the West). The heavy handed response set up the events that would lead to a real insurgency in the city over the following year and a half. Then, with the election in Iraq looming, the US could not allow what had now become a high profile insurgency in Fallujah to continue. It was decided to quell the 'City of Mosques' with an all-out assault of overwhelming military might, causing massive damage to the city and its many mosques. This action led to a huge counter-productive backlash. Its effect in Iraq was immediate when a Kurdish commander deserted after receiving a briefing about the US plans to storm the city. He was the leader of 160 Iraqi soldiers who had been training with US troops and was therefore one of the most valuable commodities that the coalition had – proof that Americans and Iraqis could work together towards a common goal.[24] Beyond Iraq, the assault generated tens of thousands of new converts to the insurgency and wider anti-Western forces. The West with its overwhelming firepower was, once again, portrayed as bombarding innocents, this time in the City of Mosques, while a brave handful of fighters stood up to the imperialist forces. Although mostly unreported in the mainstream Western media, this had the similar effect on Muslim world opinion as the destruction of something like Canterbury would have on Anglicans around the world. Even if in some strange parallel universe, it could be claimed that this was a military victory, strategically, it was a complete disaster because it did nothing but undermine the political situation at the 'big picture' level.

The US now acknowledges that the Iraqi campaign will not be won by military means alone, but will have to be a long term project linked to political and economic development.[25] However, it is clear that the temptation to use the big stick and 'take the gloves off' is sometimes pushing the military effort in a counter-productive direction. Mark 77 firebombs or 'new napalm' are being used with no sense that the

23 'Troops Kill Anti-US Protestors; Accounts Differ; 13 Dead, Many Hurt, Iraqis Say', *The Washington Post*, 30 April 2003.

24 'Make peace, not war; Brute force is no way to nurture democracy', *The Herald*, Glasgow, 8 Nov. 2004.

25 Ironically, precisely the understanding reflected in the Marine Corps Small Wars Manual in 1940.

lessons of the past have been understood at all.[26] With chilling echoes from Vietnam, the US Vice President asserted that the 'Insurgency is in its last throws' and was then flatly contradicted by the commander on the ground, General John Abizaid, who told the Senate Armed Forces Committee that 'I believe there are more foreign fighters coming to Iraq than there were 6 months ago.'[27] At the same time, and amidst the growing number of car bombs and suicide bombings, the majority of the US public now think the war was a mistake and want their troops back home.

Torture and Intelligence

A final area to consider is that of sound intelligence and how it is obtained. There is no doubt that a successful COIN campaign requires effective intelligence so that the different military, diplomatic, political and economic tools can be targeted successfully. Many have argued that the situation we now find ourselves in is such a threat to the very survival of the state, that extreme measures such as torture are now legitimate tools in the War on Terror. Leaving aside the convincing arguments that torture is an inefficient and often misleading tool that corrodes the morality, humanity and professionalism (why bother to do finger-printing when you've got a big stick?)[28] of those individuals, organizations and societies that employ or endorse it, I shall argue here that it is also another manifestation of bad strategy as it is counter-productive in the long run.

In 1956 the Algerian National Liberation Front (FLN) began a terrorist bombing campaign in Algiers, killing many innocent civilians. From 1957, General Massu's new strategy was not necessarily to go directly after the FLN bombers but to identify and eliminate anyone associated or connected with them in any way. By the end of this period this meant a very large proportion of the population were deemed potentially hostile or collaborators. Rather than selective pick ups, whole neighbourhoods were rounded up. At first, torture was used on likely suspects, presumably with what was believed to be some success. However, the middle ground, who had provided many reliable informants and sources of information, were increasingly alienated by the methods employed meaning they were less likely to come forward to help. This set up a negative feedback loop with more torture being required to gain more intelligence, further alienating the middle ground and further polarizing society against the French cause. By the end of the Battle of Algiers, orders had been issued by General Massu to detain 24,000 people. With judges finding themselves unable to deny torture warrants, and the French Government claiming that the Geneva Conventions did not apply in this situation, 80 per cent of the men and 66 per cent of the women held

---

26 http://news.bbc.co.uk/1/hi/uk_politics/411626.stm accessed 24 June, 2005.

27 http://news.bbc.co.uk/1/hi/world/americas/4123808.stm and *Sydney Morning Herald* 24 June 2005.

28 Darius Rejali, *Torture's Dark Allure* http://archive.salon.com/opinion/feature/2004/06/21/torture_1/index_np.html accessed 21 July 2005.

were tortured.[29] What the widespread use of torture demonstrated was that French military leaders had lost the support of the population at large. Some believe that the information gained through torture was essential in winning the Battle of Algiers. Even if this is true (the facts rather than the film do not actually bare this out), the use of torture clearly lost them the war.

One of the arguments that we are increasingly hearing is that the West is fighting the 'Global War Against Terror' with one hand tied behind its back. Why should we be bound by notions of fair play and international law when it is clear that Al Qaeda, the supporters of Bin Ladin or any other fanatical terrorist organization have made it abundantly clear that they will not be bound by them? Surely it makes sense in these new and trying times, to move the boundaries to respond in a way suited to the methods of our new adversaries? This is a new type of war after all. It is interesting that WWII was also a 'new type of war' in which over 50 million people lost their lives. Defeat was too dreadful to contemplate and yet the Allies somehow managed to defeat their ruthless opponent without resorting to a routine policy of employing or relying on torture. Torture, 'interrogation protocols' or any other euphemism currently employed to describe the infliction of pain and suffering to gain information, is also counter-productive. Not only is it morally wrong, it is bad strategy. It is, of course, no defence to say that you are not doing the torture yourself but letting some third party do it 'on your behalf'. There appears to be a blatant policy of sending terrorist and insurgent suspects to countries such as Morocco, Egypt and Jordon where legal restrictions on torture (or at least effective oversight) are lacking, allowing the US and her allies to get information in a way that they cannot do on their own soil. CIA counter-intelligence chief, Vincent Cannistraro is quoted as saying that 'Egyptian jails are full of guys who are missing toenails and fingernails.'[30] This 'don't ask, don't tell' policy on torture, apart from being morally abhorrent and legally very questionable at best, also does nothing but further erode the moral high ground of the West. Events in Abu Ghraib demonstrated the way that public support can be severely undermined, while fuelling the support for anyone who demonstrates that they are willing to stand up against the corruption, hypocrisy and barbarity of the West.

'The Only One Unchanging Principle in War'

Warfare will always be a terrible affair. Any attempt to regulate, moderate or control war can never remove the violence and destruction that lies at the very heart of the concept. No-one understood this better than Clausewitz: war is 'an act of violence intended to compel our opponents to fulfil our will.' However, Clausewitz also understood that war was the continuation of policy: 'there can be no question of a purely military evaluation of a great strategic issue, nor of a *purely military* scheme

---

29 Darius Rejali, *Does Torture Work?*, http://archive.salon.com/opinion/feature/2004/06/21/torture_algiers/index_np.html accessed 21 July 2005.

30 M. Dershowitz, *Why Terrorism Works* (Yale, 2002), p. 138.

to solve it.'[31] Any military operation must contribute to the overall political and strategic end state or it cannot be considered successful in any meaningful way. The purpose of strategy is to match the means to the political ends. That is precisely why the sentiment expressed by General Patton, that the only one unchanging principle in war is to 'inflict the greatest amount of death and destruction upon the enemy in the least time possible', cannot be true in many cases and is certainly untrue for any type of counter insurgency operation.

While violence and destruction can never be removed from war, the power the military wields must be controlled and harnessed towards achieving the greater political goal. The Just War Tradition (and nearly every culture, religion or society that has engaged with these issues has come up with something that looks remarkably similar) builds a powerful moral argument for adhering to principles of restraint in war. However, the principles of proportionality and discrimination from which our contemporary laws of war flow are also a practical recognition of the fact that winning is not necessarily the same as destroying your opponent. The way that you conduct your conflict has a direct effect on the outcome, which is precisely why the Russians will never win in any meaningful sense in Chechnya and why the devastating assault on Fallujah was such a counter-productive move in Iraq when it is seen from the political/strategic level.

British counter-insurgency doctrine, built on much experience (both good and bad), recognizes this and the fact that there can never be a purely military response to an insurgency. Insurgencies are defeated, not by killing everyone involved, but by persuading enough of the ordinary people that they do not want to support the insurgents. 'Hearts and minds' is not just an expression but a recognition that the people must be won over. The question of legitimacy – gaining it, maintaining it, demonstrating it and increasing it – is essential for this strategy to work. Adhering to the principles of proportionality and discrimination and demonstrating a respect for law and justice must be at the heart of any counter-insurgency strategy if it is to be effective. Clearly barbarity and legitimacy do not go together which is why torture can never be a just tool to employ and why relying upon it will simply undermine your strategy in the long run. Political success in a local insurgency all the way up to the Global War on Terror requires the cooperation and support of the public on both sides of the struggle involved. The implications for losing that cooperation and support are grave, both for the politicians at home and, in a rather more physical sense, the people on the ground.

If the means employed actually make attaining the political end state harder rather than easier, then that is clearly a bad strategy. That is precisely why the gloves must stay on.

---

31 Carl von Clausewitz, *Two Letters on Strategy*, Peter Paret and Daniel Moran (trs. and eds), (Carlisle, Penn., 1984), p. 9.

Chapter 9

The Discursive Construction of Torture in the War on Terror: Narratives of Danger and Evil

Richard Jackson

The deeply disturbing torture and murder of terrorist suspects in the war on terrorism, as well as the conspicuous cruel, inhumane, and degrading treatment by Coalition forces in Iraq and Afghanistan, is by now extremely well documented. Official reports from the US military, the International Committee of the Red Cross (ICRC), and human rights groups such as Human Rights Watch confirm that prisoners in American detention centres in Cuba, Afghanistan, Iraq, and elsewhere have been the victims of: casual and serious beatings which in some cases resulted in death, including with a broomstick and pistol-whipping; prolonged hooding; transportation in unventilated boxes and in painful restraints; sleep deprivation; sensory deprivation by exposure to bright lights and loud music; threats of removal to countries where they are likely to face torture and death; prolonged restraint in so-called 'stress' positions; being kept naked and bound for days at a time; prolonged solitary confinement; the denial of medical treatment, including to a prisoner who had been shot; the pouring of phosphoric acid over prisoners' genitals; the unleashing of attack dogs on naked prisoners; pouring cold water on naked detainees; 'water boarding' in which prisoners are forcibly submerged under water and made to believe they will drown; the sodomy of prisoners with chemical lights and broomsticks; rape and its threat; other forms of sexual humiliation and ritual domination; the mutilation of corpses; and murder (dozens of suspicious deaths in custody are presently under investigation) – among others.[1]

---

*   This study extends the research presented in my most recent book, Richard Jackson, *Writing the War on Terrorism: Language, Politics and Counter-Terrorism*, (Manchester, 2005).

1   See 'The Depositions: The Prisoners Speak, Sworn Statements by Abu Grhaib Detainees'; 'Report of the International Committee of the Red Cross (ICRC) on the Treatment by the Coalition Forces of Prisoners of War and Other Protected Persons by the Geneva Conventions in Iraq During Arrest, Internment and Interrogation', February 2004; 'The Taguba Report', March 2004; 'The Mikolashek Report', July 2004; 'The Schlesinger Report', August 2004; 'Vice Admiral Albert Church III's Brief on Investigation into Allegations of

Importantly, the public record clearly demonstrates that these specific abuses were not isolated incidents; rather, it reveals that such abuses were widespread and systematically applied. Official military investigations alone recorded around 300 official allegations of abusive behaviour towards detainees in more than 20 US-run detention centres by August 2004.[2] The ICRC made over 200 allegations of ill-treatment of prisoners of war in May 2003; they made another 50 allegations regarding ill-treatment at Camp Cropper alone in July 2003.[3] And human rights researchers have tracked more than 330 accusations of abuse against more than 460 detainees since 2001, involving more than 600 US military and civilian personnel.[4] The catalogue of abuse revealed in these and other reports has been corroborated in a number of subsequent legal trials. Furthermore, widespread allegations of human rights abuses have not been confined to American forces; British, Danish, Iraqi, and Afghan forces have all been accused of abusing detainees, and a number of British and Danish soldiers have since gone on trial for serious crimes against prisoners.

Journalistic and legal investigations further corroborate the official and non-governmental findings,[5] but also reveal a great many other instances of abusive behaviour and human rights violations unacknowledged by administration officials or purportedly still under official investigation. For example, information has emerged that under America's rendition programme, an unknown number of terrorist suspects have been secretly transported to countries where legal and ethical restraints on torture are routinely flouted, such as Jordan, Egypt, Morocco, and Syria; often under the direction of American intelligence agents, prisoners have suffered months of severe torture followed in some cases by execution or disappearance.[6] This is the long-running practice by American intelligence of torture by proxy,[7] and it is banned under international law. As a consequence of official failure to fully investigate

---

Abuse of Prisoners at Guantánamo Bay', May, 2004; and the 'Fay/Jones Report', August 2004 – among others. These documents have been published in Mark Danner, *Torture and Truth: America, Abu Ghraib, and the War on Terror*, (New York, 2004), and Karen Greenberg and Joshua Dratel, (eds), *The Torture Papers: The Road to Abu Ghraib*, (New York, 2005).

2   See 'The Schlesinger Report', August 2004, in Greenberg and Dratel, *The Torture Papers*, p. 909.

3   Reed Brody, 'The Road to Abu Ghraib', in Rachel Meerpol, (ed.), *America's Disappeared: Secret Imprisonment, Detainees, and the 'War on Terror'*, (New York, 2005), p. 123.

4   'Report: Detainee abuse claims not investigated in full', CNN.com, 26 April 2006, URL: http://www.cnn.com/2006/US/04/26/detainee.report/index.html.

5   See David Rose, *Guantánamo: America's War on Human Rights*, (London, 2004); Seymour Hersh, *Chain of Command: The Road from 9/11 to Abu Ghraib* (London, 2004); David Cole, *Enemy Aliens: Double Standards and Constitutional Freedoms in the War on Terrorism*, (New York, 2003); Meerpol, *America's Disappeared* – among others.

6   See Steven Watt, 'Torture, "Stress and Duress", and Rendition as Counter-Terrorism Tools', in Meerpol, *America's Disappeared*.

7   In testimony to the 9/11 Commission, George Tenet, then director of the CIA, stated that in an unspecified period *before* 11 September 2001, America had undertaken more than 70 such renditions. Ibid, pp. 83-4.

numerous and continual allegations of torture and abusive treatment, an American Bar Association report concluded that: 'The American public still has not been adequately informed of the extent to which prisoners have been abused, tortured, or rendered to foreign governments which are known to abuse and torture prisoners.'[8] Similarly, the Detainee Abuse and Accountability Project argued: 'It has become clear that the problem of torture and other abuse by US personnel abroad was far more pervasive than the Abu Ghraib photos revealed – extending to numerous US detention facilities in Afghanistan, Iraq, and at Guantanamo Bay, and including hundreds of incidents of abuse.'[9] It seems highly likely, therefore, that the public revelations of abuse thus far represent only the tip of the iceberg, and many more abuses go unreported.

The public record also confirms that Amnesty International, Human Rights Watch, and other human rights organizations first began to raise their concerns about the treatment of terrorist suspects swept up after 11 September 2001, as well as the treatment of captives from Operation Enduring Freedom, in early 2002 – two years before the Abu Ghraib photographs were first published. In other words, allegations of the abuse of prisoners are not without precedent; they have been reported, since the war on terror first began. In fact, investigations by journalists reveal that a policy of coercive and abusive interrogation of terrorist suspects was decided on from the very first days of the campaign against the Taliban in Afghanistan.[10] Emerging out of sustained and considered internal debate between the White House, the Pentagon, the State Department, and the Department of Justice, officials clearly acknowledged their unlawful intentions at a very early stage and instructed legal advisors to construct a robust defence of the policy to avoid potential prosecution for war crimes in the future.[11] The notion that policy-makers were merely misguided or mistaken in their eagerness to protect Americans, or the individuals who put the policy into practice were just a few ill-disciplined and sadistic loners – the 'bad apples' theory – is wholly unsupported by the publicly available evidence. The abuses took place in the context of what appears to have been a deliberate policy systematically applied over several years.

This chapter seeks to better understand the ways in which the torture policy was both formulated by the political elite – against the advice of a great many experts and respected figures such as Colin Powell – and enacted by individual torturers. More specifically, it explores the process by which such morally repugnant behaviour came to be accepted and normalized in military practice and the consciousness of the wider public. The central argument advanced here is that the practice of torture by Coalition forces is intimately connected to the social and political construction of the overall war on terrorism after 11 September 2001; it is the culmination of a

---

8   'American Bar Association Report to the House of Delegates', August 2004, in Greenberg and Dratel, *The Torture Papers*, p. 1134.

9   'Report: Detainee abuse claims not investigated in full'.

10   See Hersh, *Chain of Command*, pp. 49-50.

11   Joshua Dratel, 'The Legal Narrative', in Greenberg and Dratel, *The Torture Papers*, p. xxi.

particular kind of discursive process. Specifically, the rhetorical construction of a terrorist threat that was unprecedented and supremely catastrophic, combined with the demonization and dehumanization of the terrorist 'other', created the conditions within which torture and abuse became normalized. It is possible to demonstrate the discursive process by which this took place, in part by employing the methodology of critical discourse analysis (CDA).[12] An examination of more than a hundred administration speeches, interviews, memos, and official reports clearly demonstrates that the public language of counter-terrorism by senior administration officials was reproduced and deployed in the internal debate and policy documents relating to prisoner interrogation. Far from being simply public rhetoric or propaganda, the hyperbolic language of the new catastrophic threat of terrorism and the evil, faceless terrorist enemy actually informed and structured the real-world policies of the war on terror. Additionally, an examination of some specific instances of prisoner abuse – in particular, those portrayed in the Abu Ghraib photographs – reveal the ways in which the administration's public discourse about terrorists was reflected in the behaviour of individual soldiers towards terrorist suspects.

The chapter is divided into three sections. In the first section, I examine some broad frameworks for explaining the torture and abuse of detainees in Iraq and elsewhere. The second section describes the primary discursive processes implicated in the construction and reproduction of torture, and assesses the evidence linking the public political discourse of the war on terror to the formulation of the torture policy as well as its practice in Abu Ghraib. In the concluding section, I reflect on the analytical and normative implications of the study.

Explanations of Torture and Abuse in the War on Terror

At the very broadest level, it can be argued that the abusive violence of America's counter-insurgency campaign in Iraq and Afghanistan is simply a reflection of the inbuilt tendency towards unrestricted violence that is inherent in war generally. War is intrinsically degenerate, regardless of which form it takes;[13] as Clausewitz put it,

---

12 CDA is at once both a technique for analyzing specific texts or speech acts, and a way of understanding the relationship between discourse and social and political phenomena. By engaging in concrete, linguistic textual analysis – that is, by doing systematic analyses of spoken and written language – CDA aims to shed light on the links between texts and societal practices and structures, or, the linguistic-discursive dimension of social action. In addition, because individual text analysis is not sufficient on its own to shed light on the relationship between discourse and social processes, CDA adds a wider inter-disciplinary perspective which combines textual and social-political analysis. Further explanation of my methodology and the documents I examined can be found in Jackson, *Writing the War on Terrorism*. A thorough explanation of the CDA approach to research can be found in Marianne Jorgensen and Louise Phillips, *Discourse Analysis as Theory and Method*, (London, 2002).

13 Alex de Waal, 'Wars in Africa', in Mary Kaldor, (ed.), *Global Insecurity: Restructuring the Global Military Sector, Volume III*, (London, 2000), pp. 32-3. See also, Martin Van Creveld, *The Transformation of War*, (New York, 1991), p. 48.

escalation is a law of war and the violence of war has 'no logical limit'.[14] Directly related to this, it is well-known that violence is inherently imitative or mimetic – an act of violence is reflexively met by the opponent with a similar or greater act of counter-violence.[15] This is part of the wider psychological context of violent conflict and partly explains how human rights abuses occur and why they are ubiquitous in war; they are part of an in-built escalatory violence-counter-violence cycle. Certainly, the war in Iraq fits the typical pattern of a mimetic war of violent abuses and accompanying violent images: the photographs of orange jump-suited prisoners in Coalition prisons are discursively mirrored in the video footage of orange-suited Western captives in make-shift insurgent cages; the pictures of terrified Iraqi prisoners being savaged by dogs are mirrored by the images of Japanese hostages being threatened by knife-wielding insurgents; and so on. In essence, war is by nature violent, cruel, and inhumane; it has always been so and it is naïve or disingenuous to suggest it could be otherwise.

Within this broader context however, a more obvious explanation for the widespread abuse of prisoners in the war on terror is that, despite official protestations to the contrary, it was the direct result of a deliberate policy decision at the highest level and the abuses committed by individual soldiers were the consequence of instructions received from their superiors. As suggested above, there is in fact a great deal of evidence to support this view. In the first instance, journalistic investigations reveal that in the immediate aftermath of the 11 September 2001 terrorist attacks, senior administration officials agreed that fighting al Qaeda would involve a no-holds-barred approach which would involve employing distasteful methods, such as using foreign intelligence services to gain confessions by torture if necessary.[16] Consequently, President Bush signed a top-secret finding in late 2001 or early 2002 authorizing the Defense Department to set up a specially recruited clandestine team of operatives to snatch or assassinate 'high value' al Qaeda operatives anywhere in the world who would then be interrogated secretly and in ways unconstrained by legal limits or public disclosure.[17] This was the origin of the administration's rendition program, in which hundreds of terrorist suspects have so far been captured or kidnapped in dozens of different countries and transported to Guantánamo Bay or to third countries like Syria, Jordan, Egypt, or Singapore for interrogation. This initial authorization for a secret 'dirty war' against terrorists was quickly institutionalized as an 'unacknowledged' special-access program (SAP) in the Defense Department known inside the intelligence community as 'Copper Green'.[18]

The important point is that this secret global program in which interrogators were authorized to use methods that went beyond legal limits was later extended to

---

14 Quoted in Martin Shaw, *War and Genocide: Organized Killing in Modern Society*, (Cambridge, 2003), p. 19.

15 The military historian, Martin Van Creveld, suggests that 'war represents perhaps the most imitative activity known to man (sic)'. See Van Creveld, *The Transformation of War*, p. 174 and p. 195.

16 See Bob Woodward, *Bush at War*, (London, 2003), pp. 74-8.

17 Hersh, *Chain of Command*, p. 16.

18 Ibid, p. 46.

Iraq as a means of combating the growing insurgency there. It was in this context that individual prison guards received instructions from intelligence operatives to 'set favourable conditions for subsequent interviews'[19] – a euphemism for breaking the will of prisoners, or 'softening them up'. An official military investigation into prisoner abuse in Iraq acknowledged that: 'There was a perception among the guard personnel that this type of behaviour by the interrogators was condoned by their chain of command.'[20] Similarly, the Taguba Report concluded that 'personnel assigned to the 372nd MP Company, 800th MP Brigade were directed to change facility procedures to "set the conditions" for MI [Military Intelligence] interrogations.'[21] In other words, it seems highly likely that it had already been decided at a senior level to extend the use of coercive and abusive interrogation methods – 'any means necessary' in the language of the presidential finding – to all theatres of the war against terrorism, and that prison guards were encouraged to assist the process by using cruel, inhumane, and degrading treatment to psychologically prepare the prisoners for questioning. Certainly, the Abu Ghraib photographs, with their confident smiles and bustling background activity, would seem to indicate that the perpetrators did not fear any serious consequences for the behaviour. The fact that the photographs were subsequently swapped from computer to computer throughout the 320th Battalion – being used as a screensaver in at least one instance – further indicates that they felt no need to hide their actions.

Furthermore, the argument that the abuse was carefully calculated for strategic reasons, rather than being random or pointless sadism, is clearly demonstrable. There is an extremely large literature detailing the strategic logic of employing torture and conspicuous human rights abuses in asymmetric war[22] – as a means of preventing civilian defection to the enemy, for example. More specifically, there is evidence that the sexual humiliation seen in the Abu Ghraib photographs was deliberately conceived as a culturally-specific method of breaking down the psychological resistance of Iraqi detainees; military and intelligence officials believed that Arabs were particularly vulnerable to sexual humiliation due to cultural conditioning.[23] In

---

19 'The Ryder Report', quoted in Hersh, *Chain of Command*, p. 28. See also Rose, *Guantanamo*, pp. 87-8.

20 'The Mikolashek Report', July 2004, in Greenberg and Dratel, *The Torture Papers*, p. 655.

21 'The Taguba Report', March 2004, in Danner, *Torture and Truth*, p. 294.

22 See among many others, Roger Beaumont, 'Small Wars: Definitions and Dimensions', *Annals of the American Academy of Political and Social Sciences*, 541 (Small Wars), 1995, pp. 20-35; Stathis Kalyvas, 'Wanton and Senseless? The Logic of Massacres in Algeria', *Rationality and Society*, 11, 1999, pp. 243-85; M.L.R. Smith, 'Guerillas in the Mist: Reassessing Strategy and Low Intensity Warfare', *Review of International Studies*, 29, 2003, pp. 19-37; Paul Richards, *Fighting for the Rain Forest: War, Youth, and Resources in Sierra Leone*, (Oxford, 1996).

23 Various commentators note that senior administration officials were greatly influenced by *The Arab Mind*, a book on Arab culture and society published by Raphael Patai in 1973. The book includes a long chapter on Arab attitudes towards sex, concluding that sexual

addition to their role as trophies and the added humiliation for the detainees of being photographed, the evidence also suggests that photographing the abuses was itself a strategic decision: it was thought that some prisoners could be induced to spy on their associates to avoid dissemination of the shameful photographs to their friends and family.[24] From this perspective, it could be argued that Coalition forces (and the insurgents who engaged in broadly imitative behaviour) were simply adhering to Clausewitzean nostrums regarding the instrumental rationality of force by using all means necessary to try to bend the enemy to their will.[25]

Lastly, the argument that it was a deliberate and calculated policy is buttressed by the evidence contained in the paper trail of the administration's internal debates over the legality of such practices. For example, the internal memos between the Justice Department and the administration reveal that senior administration officials deliberately chose Guantánamo as a site for prisoner confinement and interrogation specifically because it was believed to afford a degree of legal protection from the reach of American and international courts.[26] In addition, the Justice Department advice to the White House strongly indicates that the administration had already decided on the policy and was looking for legal justification, as well as legal protection, for employing methods that they knew went beyond national and international legal restrictions.[27] In other words, the administration's defence of the policy shows evidence of deliberate intention and careful forethought.

In addition to this policy-level explanation, the most frequent public explanation given by the Bush administration is that they were the actions of a very small group of ill-disciplined and morally degenerate individuals, who even if they had been instructed to set the psychological conditions for interrogation, clearly went far beyond acceptable standards of military behaviour. At one level, there is an element of truth to this claim: some of the guards at Abu Ghraib were poorly trained private security contractors and reservists, while others had poor work and discipline records. The Pentagon admitted in an internal report that many of its intelligence officers in Afghanistan for instance, were very poorly trained.[28] Similarly, military investigations into the abuses in Iraq identify poor training, along with weak leadership and oversight, as one of the key explanatory variables for the abuse of

---

subjects are a cultural taboo invested with shame in Arab society. This academic 'knowledge' was then used to fashion the sexual humiliation strategies vividly portrayed in the Abu Graib photographs. See Hersh, *Chain of Command*, pp. 38-9.

24 Ibid.

25 Smith, 'Guerillas in the Mist', p. 26.

26 According to a senior Pentagon official, Guantánamo Bay was chosen precisely because it was seen to have a low risk of litigation by virtue of its unique location and legal status. Rose, *Guantánamo*, p. 33.

27 As Anthony Lewis put it, administration lawyers 'were asked how far interrogators could go in putting pressure on prisoners to talk without making themselves, the interrogators, liable for war crimes.' Anthony Lewis, 'Introduction', in Greenberg and Dratel, *The Torture Papers*, p. xiii.

28 Rose, *Guantánamo*, p. 45.

prisoners.[29] Individual pathology and inexperience clearly does play a role in some cases of abuse, especially when vetting processes of prison guards fail, training is inadequate, and supervision is lax. However, as mentioned above, the wealth of evidence does not fully support the 'bad apples' explanation; the abuses were far too widespread, strategically conceived, and systematically administered to be the work of a few isolated individuals.

Another explanatory perspective relevant to the war on terror derives from social-psychology, particularly a number of experiments undertaken in the 1970s, as well as evidence from prisoner of war camps and military training simulations. Stanley Milgram for example, demonstrated how easy it was to persuade ordinary people to cause severe pain in others, especially when ordered to do so by an authority figure.[30] The diffusion of responsibility through multiple and overlapping layers of authority can also contribute to such abuses by magnifying this effect. The Stanford prison experiment in 1971 by psychologist Philip Zimbardo revealed that when ordinary people are given absolute power over others they very quickly become sadistic and abusive towards their captives, particularly in the presence of weak supervision regimes and the initial toleration of minor abuses. Due to escalating levels of sadism by guards, the experiment had to be closed down after just six days. Military research in war resistance training similarly demonstrates that regardless of how good the training and oversight, some inappropriate behaviour from guards will always occur in a prisoner situation.[31] In this instance, the social-psychology of group dynamics led to individuals behaving in ways that they and society would ordinarily find abhorrent. In other words, the instructions given by authority figures in the war on terror to 'set the conditions' for interrogation, combined with the group dynamics of a stressful social situation in which prison guards had absolute control over the bodies of the inmates, greatly reduced the behavioural constraints that might have prevented the abuses in Iraq and elsewhere.

The brutalization of soldiers during training plays a key role in de-sensitizing soldiers to physical and psychological violence against detainees; subject to de-personalization, uniforms, lack of privacy, lack of sleep, disorientation, punishing physical regimes, harsh and often capricious punishments, and violent hazing rituals, soldiers come to accept arbitrary and frequently sadistic violence as normal.[32]

---

29 A report in July 2004 for example, found that many prison guards did not have training specific to detainee handling, and a great many individuals involved in interrogation were not school-trained as interrogators. See 'The Mikolashek Report', 21 July 2004, in Greenberg and Dratel, *The Torture Papers*, p. 656.

30 Discussed in Joanna Bourke, *An Intimate History of Killing: Face-to-Face Killing in Twentieth-Century Warfare*, (London, 1999), p. 185.

31 'The Mikolashek Report', 21 July 2004, in Greenberg and Dratel, *The Torture Papers*, p. 653.

32 See Bourke, *An Intimate History of Killing*, p. 67. Bourke demonstrates that the methods used to train Western soldiers during World War I and II and the Vietnam War were very similar to those carried out in regimes where men were taught to torture prisoners. The difference between the two resided in the degree of violence involved, not its nature.

This is a crucial component of the military training process which is necessary for transforming ordinary people into disciplined soldiers capable of killing on command; overcoming social and psychological inhibitions to committing violence against other human beings actually requires a carefully formulated and sustained training program over several months.[33] It is more than a coincidence that many of the abuses recorded in Iraq mirror the treatment meted out to recruits as part of their training and initiation to military life. In part, this is why Lynndie England claimed that 'it was just fun, harmless fun' – nothing more than a 'prank'. Others agreed: Rush Limbaugh told his radio listeners that it was 'no different than what happens at the Skull and Bones initiation.'[34] In other words, the abuses appeared to these soldiers as little more than an elaborate hazing ritual similar to what they themselves had invariably experienced. In this sense, the Abu Ghraib abuses and regular institutional military training are discursively part of the same single culture of violence.

Another element specific to the war on terror is the ubiquitous and deliberate use of the 11 September 2001 attacks as a source of motivation for American soldiers. The military has always employed propaganda about the grievous wrongs of the enemy to inspire and motivate its soldiers; during World War II, the attack on Pearl Harbor was deliberately used to instil hatred towards the Japanese among new recruits. In the war on terror, US military recruitment and training films frequently employ images of 9/11, and posters and monuments to the attacks abound at the American base in Guantánamo Bay,[35] among others. There seems little doubt that the constant reminders of the atrocities function to sustain anger and hatred towards America's enemies, and to de-sensitize soldiers to the suffering of 'terrorist' suspects. Considering the large number of Americans who continue to believe that Iraq was involved in the terrorist attacks, hatred and anger towards Iraqi detainees by American soldiers is probably unsurprising.

While the combination of these explanatory factors – the inherently degenerate nature of asymmetric warfare, the deliberate and strategic application of a torture policy, poor training and sadistic individuals, the social-psychology of prisons, military training, and the continual reminders of 9/11 – goes some way towards explaining the occurrence of torture in Iraq and elsewhere, it leaves a number of crucial questions unanswered. In the first place, these perspectives fail to explain why the wider public and society in general acquiesce to such behaviour by their representatives – why public support for the troops and the political elite has remained high even after it became publicly known that such abuses were widespread and deliberately enacted. In addition, they do not fully explain why leaders choose to employ such tactics, even when they are aware of their strategic limitations and political dangers. The

---

33 See Dave Grossman, *On Killing: The Psychological Cost of Learning to Kill in War and Society*, (New York, 1995).

34 Quoted in Ziauddin Sardar, 'The Holiday Snaps', *Newstatesman* (7 March 2005), p. 49.

35 Rose, *Guantánamo*, p. 58.

official record reveals that a number of senior administration officials, including Secretary of State Colin Powell and his legal advisor, William H. Taft, cautioned against adopting such practices for a range of compelling reasons, including that it could undermine the legal protection of American troops.[36] Moreover, legal briefings from administration lawyers acknowledged the many arguments against adopting torture, but then ignored them because their real purpose was to establish legal protection for a policy that had already been decided.[37] At the operational level, the standard rulebook for American military interrogators, a document known as Field Manual 34-52, actually prohibits the use of coercive techniques because they produce low quality intelligence. The manual states: 'The use of force is a poor technique, as it yields unreliable results, may damage subsequent collection efforts, and can induce the source to say whatever he thinks the interrogator wants to hear.'[38] In fact, a significant number of White House officials, senior military officers (including a general who was later removed from duty at Camp Delta for expressing his concerns at the policy), FBI agents, Pentagon advisors, and other intelligence operatives expressed their dismay and opposition to the torture policy.[39] In other words, why did senior officials authorize techniques which were clearly counter-productive and which have since been extremely damaging to the prosecution of the war on terrorism?

The Discursive Construction of Torture

It is not sufficient to explain torture solely by reference to the broad political and social-psychological factors mentioned above. What actually needs to be explained is how whole societies and the individuals within them come to view an enemy 'other' as inherently dangerous, inhuman, demonic, and undeserving of even the most minimal levels of human respect. The central process in the construction and reproduction of torture and abuse, I believe, is the deliberate creation by political

---

36 See Memorandum for Counsel to the President, Assistant to the President for National Security Affairs, from Colin L. Powell, Draft Decision Memorandum for the President on the Applicability of the Geneva Convention to the Conflict in Afghanistan, 26 January 2002, in Greenberg and Dratel, *The Torture Papers*, pp. 122-25, and Memorandum for Counsel to the President, from William H. Taft, IV, Comments on Your Paper on the Geneva Convention, 2 February 2002, in Greenberg and Dratel, *The Torture Papers*, pp. 129-33.

37 See Memorandum for Alberto R. Gonzales, Counsel to the President, from Jay S. Bybee, Re: Standards of Conduct for Interrogation under 18 U.S.C., 1 August 2002, in Danner, *Torture and Truth*.

38 Quoted in Rose, *Guantánamo*, p. 95. Anthony Lewis argues that suppositions of what a suspect knows are usually wrong, and that statements extracted by torture have repeatedly been found to be unreliable. There is no reason to suspect that administration lawyers and officials were unaware of this. Lewis, 'Introduction', in Greenberg and Dratel, *The Torture Papers*, p. xvi.

39 See for example, Hersh, *Chain of Command*, pp. 6-7, 13-14, 42-3, 60-2. See also Watt, 'Torture, "Stress and Duress"', in Meerpol, *America's Disappeared*, p.76.

entrepreneurs of a totalizing and hegemonic social and political discourse, a 'vast cultural complex' that deconstructs existing discourses of tolerance and respect for human rights, and their replacement with new discourses of hatred, fear, and the justified use of extreme violence.[40] In this respect, public political discourses function to set the vocabulary and parameters of debate, control public opinion, suppress opposition, and create new collective norms of acceptable violence. This approach integrates structural and agentic understandings of political violence, and links the social-political and the social-psychological levels of human behaviour.

There are two main processes visible in the discursive construction of torture in the war on terror. In the first instance, the language and practice of counter-insurgency, domestic prison management, and the treatment of immigrants in American political life provides concrete historical continuities with the treatment of prisoners in Iraq and elsewhere. From this perspective, the prisoner abuse scandal reveals clear discursive ties with previous practices towards particular demonized groups, and could be described as reflexive. Second, the core public narratives of the war on terrorism declared after 11 September 2001 discursively construct a social and political reality in which such behaviour becomes not only possible, but highly predictable. That is, constructing the terrorist 'other' as dangerous, inhuman, and inherently evil functions to socially legitimize and ultimately normalize the practice of abusive treatment across the military and society.

The Treatment of Prisoners in Historical Context

The historical continuities in American counter-insurgency practice, domestic prisoner management, the treatment of immigrants, and the abusive behaviour towards detainees in the war on terror are striking – and more than coincidental. In reality, these practices act as a discursive learning process and go some way towards explaining the construction and reproduction of torture and abuse. Over time, the language and practice of prisoner management has become embedded in institutional knowledge and practice, standard operating procedures, and collective memory and popular culture; to this extent, the abusive treatment of prisoners by American officials is reflexive.

In the first instance, the historical experience of American counter-insurgency demonstrates clear continuities in practice from the earliest colonial times. The Seminole wars in Florida for example, were characterized by classic dirty war tactics,[41] as was the war against the Dakota Sioux in 1862. During this campaign hundreds of prisoners were declared 'unlawful combatants' and subject to summary trial by military commission; some 38 were later executed.[42] More generally, the

---

40 See Richard Jackson, 'The Social Construction of Internal War', in Richard Jackson, (ed.), *(Re)Constructing Cultures of Violence and Peace*, (Amsterdam, 2004).

41 Ian Beckett, *Modern Insurgencies and Counter-Insurgencies: Guerillas and the Opponents since 1750*, (New York, 2001), pp. 29-30.

42 Rose, *Guantánamo*, pp. 138-40.

management of captured and defeated Native Americans during counter-insurgency and resettlement operations reveals numerous continuities with the treatment of insurgents in Iraq today. Later, the period of American colonial rule in the Philippines witnessed 117 verifiable atrocities against Filipino civilians between 1898 and 1902 – at the cost of thousands of lives.[43] Similar practices characterized anti-communist counter-insurgency during the cold war. The Phoenix program in Vietnam for example, employed assassination and torture as a means of rooting out Vietcong agents, leading to the deaths of some 25,000 suspected Vietcong.[44] The Church Commission in 1975, established by the Senate Committee on Intelligence, found that in addition to the abuses in Vietnam, the CIA had been involved in comparable activities in other parts of the world, notably Latin America.[45] Interrogation approaches developed in Vietnam were subsequently exported to Latin America via the notorious School of the Americas (now called the Western Hemisphere Institute for Security Cooperation). Training manuals from this period, notably the *KUBARK Counterintelligence Interrogation Manual, July 1963*, and its successor, the *Human Resource Exploitation Training Manual, 1983*, detail very similar kinds of techniques to those currently employed in the war on terror.[46] Interestingly, John Negroponte, the new ambassador to Iraq, and James Steele, US advisor to the Iraqi security forces, both served in Central America during the 1980s; Negroponte was implicated in the activities of the Contras in Nicaragua and the death squads in Honduras.[47] During America's first war against terrorism declared by Ronald Reagan, illegal rendition, torture, and murder were employed against Middle East terrorist suspects.[48] In other words, from a historical perspective, the abuses in Iraq and elsewhere follow a long-established path in American counter-insurgency doctrine and practice. Counter-insurgency warfare has always been performed this way – such abuses are in a sense a 'normal' mode of behaviour and it would have been more surprising if these abuses had not occurred at all.

This is not to say that American counter-insurgency and counter-terrorism stands alone in its brutality. Historical studies clearly demonstrate that counter-terrorism and counter-insurgency have virtually always involved systematic and widespread human rights abuses against suspected insurgents and terrorists – even by liberal democracies. The examples are legion. For example, the use of torture

43 Beckett, *Modern Insurgencies and Counter-Insurgencies*, p. 38.

44 Ibid, pp. 201-2. See also, Bourke, *An Intimate History of Killing*.

45 See Loch Johnson, *America's Secret War: The CIA in a Democratic Society*, (Oxford, 1989).

46 See 'Prisoner Abuse: Patterns from the Past', *National Security Archive*, Electronic Briefing Book No.122, The George Washington University, URL: http://www.gwu.edu/~narchive/NSAEBB/NSAEBB122/. See also, Doug Stokes, *America's Other War: Terrorizing Colombia*, (London, 2004).

47 See Isabel Hilton, 'Torture: Who Gives the Orders?', OpenDemocracy Forum, URL: http://www.openDemocracy.net.

48 See David Wills, *The First War on Terrorism: Counter-Terrorism Policy During the Reagan Administration*, (Lanham, 2003).

and conspicuous atrocity was a feature of colonial counter-insurgency and social control policies by the Belgians in the Congo, the Germans in West Africa, the Portuguese in its African Territories, the British in Malaya, Cyprus, and Kenya, and the French in Indochina and Algeria – among many others.[49] In particular, counter-terrorism campaigns have, almost without exception, involved torture and human rights abuses by the authorities; examples include the British in Northern Ireland, Spain in the Basque region, Italy against the Red Army factions, and Israel in the Occupied Territories – to name a few. In every single case, numerous instances of coercive and abusive forms of prisoner interrogation were documented. Moreover, strategies and doctrines were discursively diffused from one location to another: American counter-insurgency in Vietnam borrowed heavily from British counter-insurgency in Malaya, for example; British methods in Northern Ireland were taken from the Cyprus campaign. Interestingly, American counter-insurgency tactics in Iraq have been greatly influenced by Israeli practices in the Occupied Territories through training by Israeli military instructors.

Second, the abuses against prisoners in Iraq follow well-established and long-term practices in domestic approaches to prisoner management, particularly in American supermaximum or 'supermax' prisons which are designed for the most dangerous felons. Amnesty International reports that more than 20,000 prisoners are currently held in more than 40 supermax prisons across America in conditions of long-term social isolation, extreme sensory deprivation, permanently lit cells, highly restricted exercise, severe forms of shackling, harsh discipline, and the like.[50] Journalistic investigations also reveal widespread and systematic abuse of prisoners in the wider American penal system, with inhumane and dangerous forms of manacling, excessive use of chemical agents to punish and subdue prisoners, severe punishment regimes, and the like. It thus appears that many of the prisoner management practices developed in America's domestic prisons have been transferred to the operating practices of the war on terror.[51] In fact, in many cases, prison guards in Iraq have had previous experience in the domestic penal system; Charles Graner for example, one of the Abu Ghraib torturers, was formerly employed at the maximum security Greene Correctional Facility in Pennsylvania; Ivan Frederick, another of the Abu Ghraib torturers, worked for the Virginia State prison system for seven years.

Lastly, the way that immigrants have been treated by the authorities also reveals striking discursive continuities with the war on terror, particularly after the 11 September 2001 attacks when thousands were caught up in a general sweep by the authorities. Interestingly, prior to the terrorist attacks, Guantánamo Bay was

---

49 Colonial authorities killed over 100,000 people in German East Africa and German South West Africa for example, and French troops tortured and killed tens of thousands of suspected FLN insurgents during the Algerian independence struggle. See Beckett, *Modern Insurgencies and Counter-Insurgencies*, pp. 24-51.

50 Rachel Meerpol, 'The Post-9/11 Terrorism Investigation and Immigration Detention', in Meerpol, *America's Disappeared*, pp. 149-50.

51 See James Meek, 'People the Law Forgot', *The Guardian* (3 December 2003), URL: http://www.guardian.co.uk/print/0,3858,4810625-111575,00.html.

used as a detention centre for Haitian and Cuban refugees; they too were held for extended lengths of time without access to lawyers or judicial processes, often in appalling conditions. As with the Camp Delta detainees, the site was chosen because it was considered to be beyond the reach of domestic courts.[52] Since the terrorist attacks however, evidence has emerged of the frequent mistreatment of immigration detainees. This has been documented in official inquiries, as well as through legal and journalistic investigations. On any given day, over 20,000 'immigration detainees' languish in American federal, state, county, and private prisons, often in deplorable and inhuman conditions similar to those reserved for supermax prisoners, and often without access to legal counsel.[53] According to an Office of the Inspector General (OIG) report in December 2003, post-9/11 immigration detainees have been the systematic and frequent victims of physical brutalities such as: slamming, bouncing, and ramming detainees against walls; bending detainees' arms, hands, wrists, and fingers; pulling and stepping on detainees' restraints to cause pain; the improper use of restraints; and rough and inappropriate handling.[54] This is in addition to the psychological suffering induced by social isolation, indefinite detainment, lack of legal representation, and the like.

Related to these historical experiences, American prisoner interrogation and management approaches have always employed a sanitized language that functions to obscure the physical and psychological effects of the material practices on the bodies of detainees. Just like the specialized military language of 'collateral damage' and 'surgical strikes', the interrogator's euphemistic language of 'counter-resistance strategies', 'stress and duress' techniques, 'non-injurious physical contact', 'stress positions', 'forced grooming', 'ego down techniques', 'rendition', and the like, functions to emotionally distance both the interrogator and the wider public from the human pain and suffering involved in these practices.

In sum, from a long-term and broad spectrum perspective, the abuses in Iraq come as no surprise. Across all American security institutions, from the prison service to the immigration service, to the CIA and the military, the harsh and often brutal treatment of prisoners is standard practice and not at all uncommon. Not only this, it has been the norm for more than a hundred years. In this sense, the discursive foundations for the abuses of the war on terror were firmly established long before 11 September 2001; the abuse of terrorist suspects was reflexive of institutional culture.

Narratives of Evil Terrorists and Good Americans

In addition to the institutionalized and embedded patterns of prisoner treatment, after the terrorist atrocities, senior administration officials constructed a totalizing and powerful discourse that not only amplified the sense of exceptional grievance

---

52 Michael Ratner, 'The Guantánamo Prisoners', in Meerpol, *America's Disappeared*.

53 Meerpol, 'The Post-9/11 Terrorism Investigation', in Meerpol, *America's Disappeared*, p. 144.

54 Ibid, p. 153.

across American society, but also thoroughly demonized and dehumanized the enemy 'other' whilst simultaneously justifying the employment of massive counter-violence. The construction and amplification of the overall public discourse of the war on terror has been explored in some detail elsewhere.[55] In terms of the prisoner abuse scandal, one of the most important aspects of the public discourse was the ubiquitous trope of the 'evil' terrorist 'other', discursively linked to the civilization/barbarism narrative. President Bush in particular, in virtually every post-9/11 speech about terrorism, suggested that America was 'in a conflict between *good and evil*',[56] and that '*evil is real*, and it must be opposed.'[57] Deeply embedded in American rhetorical traditions and religious life, this kind of language functioned to essentialize the terrorists as satanic and morally corrupt. On the day of the terrorist attacks, Bush stated that: 'Today, our nation saw *evil*, the very worst of human nature.'[58] Critically, this text locates evil in the human nature of the terrorists, thereby stigmatizing a whole category of people. In subsequent texts, senior administration officials even referred to terrorists as 'the evil ones' and 'evildoers'; these are explicitly theological constructions which equate terrorists with the Devil. In this agent/act ratio, the character of the terrorists precedes their actions: the terrorists did what they did because it is in their nature to do so – they murdered because that is what evil, demonic terrorists do.[59] In addition, this language inadvertently supernaturalizes the terrorists, drawing on cinematic depictions of evil and amplifying the threat they pose, particularly to the moral community.

It is a powerful discourse and an act of demagoguery which functions to de-contextualize and de-historicize the actions of terrorists and insurgents, emptying them of any political content, while simultaneously de-humanizing and de-personalizing them. After all, there can be no deeper explanation for 'acts of evil', and there can be no reasoning or compromising with evildoers; the only right response is exorcism and purification. Another primary effect of this discourse is to justify eradication,

---

55 See Jackson, *Writing the War on Terrorism*; John Collins and Ross Glover, (eds), *Collateral Language: A User's Guide to America's New War*, (New York, 2002); Sandra Silberstein, *War of Words: Language, Politics and 9/11*, (London, 2002); and John Murphy, '"Our mission and our moment": George W. Bush and September 11', *Rhetoric and Public Affairs*, 6, 2003, pp. 607-32.

56 George W. Bush Jr, 'Remarks by the President at the 2002 Graduation Exercise of the United States Military Academy', West Point, New York, 1 June 2002, URL: http://usinfo.state.gov/topical/pol/terror/, accessed 29 August 2003. All emphasis in direct quotations from administration officials in this chapter has been added by the author. Quotations from administration speeches are indicative samples only, reflecting literally hundreds of similar examples found by the author.

57 George W. Bush Jr, 'State of the Union Address', 29 January 2002, URL: http://usinfo.state.gov/topical/pol/terror/, accessed 29 August 2003.

58 George W. Bush Jr, 'Statement by President George W. Bush, Address to the Nation', 11 September 2001, URL: http://usinfo.state.gov/topical/pol/terror/, accessed 29 August 2003.

59 See Murphy, '"Our mission and our moment"', p. 616.

harsh treatment, and the suspension of legal protections, as well as to cauterize the possibility of sympathy or empathy for those suspected of involvement in terrorism. In other words, believing that the detainees were by nature evil and demonic sub-humans rather than ordinary people, it is not difficult to see how abuses by American soldiers became normalized in daily practice.

On the other side of this discursive coin, the American 'self' was simultaneously constructed as essentially good and heroic. A major discursive inscription of the American character came at the Prayer and Remembrance Day service on 14 September 2001. At this symbolically charged and powerfully constitutive pageant, Bush stated:

> In this trial, we have been reminded, and the world has seen, that our fellow Americans are *generous* and *kind*, *resourceful* and *brave*. We see our national character in rescuers working past exhaustion; in long lines of blood donors; in thousands of citizens who have asked to work and serve in any way possible. And we have seen our national character in eloquent acts of *sacrifice*. [...] In these acts, and in many others, Americans showed a deep commitment to one another, and an abiding *love* for our country. Today, we feel what Franklin Roosevelt called the warm courage of national *unity*.[60]

In this speech, Bush constructs a world of clearly demarcated moral identities: where terrorists are cruel, 'the American people' are generous and kind; where terrorists are hateful, Americans are loving; where terrorists are cowardly, Americans are brave and heroic; and where terrorists hide and run, Americans are united.

A related motif in the administration's public discourse was the 'hero' narrative; modelled on popular entertainment scripts, this language is very familiar in American political discourse. In the public narrative of the war on terrorism, it appeared as if every EMS worker was like Bruce Willis in *Die Hard*, every member of the armed forces was like Tom Hanks in *Saving Private Ryan*, and every ordinary citizen was like Mel Gibson in *The Patriot*. Donald Rumsfeld, in a memorial service for the Pentagon victims, constructed these all-American heroes thus:

> We remember them as *heroes*. And we are right to do so. [...] 'He was a *hero* long before the eleventh of September,' said a friend of one of those we have lost – 'a *hero* every single day, a *hero* to his family, to his friends and to his professional peers.' [...] About him and those who served with him, his wife said: 'It's not just when a plane hits their building. They are *heroes* every day.' '*Heroes* every day.' We are here to affirm that.[61]

In its discursive function, this rhetoric works to inscribe the American people, in particular American soldiers, with certain mythical qualities. In turn, the qualities

---

60 George W. Bush Jr, 'President's Remarks at National Day of Prayer and Remembrance, the National Cathedral', 14 September 2001, URL: http://usinfo.state.gov/topical/pol/terror/, accessed 29 August 2003.

61 Donald H. Rumsfeld, Secretary of Defense, 'Remarks at a Memorial Service in Remembrance of Those Lost on September 11th', The Pentagon, Arlington, VA, Thursday, 11 October 2001, URL: http://usinfo.state.gov/topical/pol/terror/, accessed 29 August 2003.

of heroism provide the military with a moral latitude that is unavailable to ordinary people; heroes are above criticism and moral judgement precisely because they are heroes; they are allowed to use extreme violence, even if it looks identical to the violence of 'evildoers'.

In the end, the administration's discursive construction of the national character had a corrosive effect on the moral community by framing American actions within a consequentialist ethical framework, and by removing the necessity for continual self-reflection and oversight. Just as the heroes in American action movies can be forgiven excessive brutality because they are ultimately fighting on the side of good, so too the American military can be forgiven any 'collateral damage' in the noble cause of liberty and justice. In other words, if the eventual outcome of the war can be said to result in victory over terrorism and the establishment of 'enduring freedom', abuses along the way can be weighed up against the greater good that was achieved. A sense of moral superiority by policy-makers leads to the mistaken belief that regulation and oversight is unnecessary because good people (soldiers) instinctually produce good behaviour.

These primary narratives of 'self' and 'other' identities were established across American society and consequently within the American military, largely through their constant repetition, their relatively unmediated transmission to the public via the mainstream media, and their amplification through other social institutions like churches, schools, universities, and the like. Senior administration officials have given more than six thousand speeches, interviews, radio broadcasts, and the like, on the war against terrorism since 11 September 2001, averaging around ten per day over the entire period.[62] In the first 26 days after 11 September to 6 October, President Bush alone gave more than 50 public statements on the war on terrorism, most with extensive media coverage; and in the months from 7 October to 7 December, he made 76 public statements, using opportunities such as the visit to an elementary school or the joint appearance with President Putin of Russia to relay his message.[63] In all these speeches, the primary narratives of American grievance, evil terrorists, heroic American soldiers, and the good war against terrorism were reiterated time and again. In essence, reproduced across American society in an almost endless cycle, the effect was a kind of discursive saturation of American society – a blitzkrieg on public and private consciousness – which functioned to set the parameters of all public and private debate, suppress doubts about the government's policies, and establish new norms of ethical behaviour.

The discourse was then reinforced by a series of publicized actions by the administration that powerfully confirmed the evil and inhuman status of the enemy terrorist 'other'. For example, during Operation Enduring Freedom in Afghanistan, Donald Rumsfeld stated quite openly that he would prefer it if Taliban and al Qaeda fighters were killed rather than be allowed to surrender; the Special Forces operating

---

62 Jackson, *Writing the War on Terrorism*, p. 163.

63 Brigitte Nacos, *Mass-Mediated Terrorism: The Central Role of the Media in Terrorism and Counterterrorism*, (New York, 2002), pp. 148-9.

in Afghanistan were given authority to kill on sight. Combined with the Military Order of 13 November 2001 in which combatants in Afghanistan were denied protection under the Geneva Conventions, these actions sent a powerful message to troops in the field about how the lives of the enemy should be regarded. The American Government also arrested thousands of suspects after 11 September 2001, denying them even the most basic of civil and legal rights. Again, this action clearly demonstrated that the enemy 'other' was to be treated outside the parameters of normal jurisprudence and human rights standards.

The Public Discourse-Policy Formulation Connection

Thus far, I have argued that the combination of existing historically embedded practices and understandings of prisoner management and interrogation, combined with a totalizing public discourse of evil and demonic terrorists versus good Americans, made torture and abuse highly likely. Further evidence for this thesis can be observed in the way that a number of other core public narratives about the war on terrorism infiltrated the formulation and defence of the torture policy. For example, the administration's public rhetoric discursively constructed the terrorist attacks as 'acts of war', thereby invoking the nation's right to 'justified self-defence' based on international law. President Bush asserted that '*war has been waged* against us,'[64] and 'the wreckage of New York City' was 'the first *battle of war*.'[65] Under Secretary of State Marc Grossman went on to argue that 'we believe the United States was attacked on the 11th of September and that we have *a right of self-defence* in this regard.'[66] Donald Rumsfeld appealed directly to the universal right of every nation to self-defence: 'there is no question but that any nation on Earth has *the right of self-defence*. And we do.'[67] Interestingly, the adoption of a war-based rhetorical mode did not occur immediately;[68] for the first few days senior administration officials described them as 'deliberate and deadly *terrorist acts*' and 'despicable *acts of terror*'.[69] Of course, this language was not inadvertent or mere exaggeration; strategically, declaring a state of war invests the state with powers that are unavailable during peacetime.

---

64 Bush Jr, 'President's Remarks at National Day of Prayer and Remembrance', 14 September 2001.

65 George W. Bush Jr, Colin Powell, and John Ashcroft, 'Remarks at Camp David', 15 September 2001, URL: http://usinfo.state.gov/topical/pol/terror/, accessed 29 August 2003.

66 Marc Grossman, 'Interview of Under Secretary of State', Digital Video Conference, 19 October 2001, Washington, DC, A trans-Atlantic digital interview with London-based journalists of Arab newspapers, URL: http://usinfo.state.gov/topical/pol/terror/, accessed 29 August 2003.

67 Donald H. Rumsfeld, Secretary of Defense, 'Interview with Wolf Blitzer, CNN', 28 October 2001, URL: http://usinfo.state.gov/topical/pol/terror/, accessed 29 August 2003.

68 Jackson, *Writing the War on Terrorism*, pp. 38-40.

69 Bush Jr, 'Address to the Nation', 11 September 2001.

This public discursive construction of the state of conflict – that the terrorist attacks constituted a state of war – immediately found its way into the key documents of the torture policy. The President's Military Order of 13 November 2001 for example, which denied captured al Qaeda and Taliban fighters protection under the Geneva Conventions stated: 'International terrorists, including members of al Qaida, have carried out attacks ... on a scale that has created *a state of armed conflict* that requires the use of the United States Armed Forces.'[70] Similarly, legal advice from the Attorney General's office stated:

> As we have made clear in other opinions involving *the war* against al Qaeda, *the nation's right to self-defense* has been triggered by the events of September 11. If a government defendant were to harm an enemy combatant during an interrogation in a manner that might arguably violate Section 2340A, he would be doing so in order to prevent further attacks on the United States by the al Qaeda terrorist network. In that case, we believe that he could argue that his actions were justified by the Executive branch's constitutional authority *to protect the nation* from attack. This national and international version of *the right to self-defense* could supplement and bolster the government defendant's individual right.[71]

In this text we can observe a direct link drawn between the discursive construction of the 11 September 2001 attacks as acts of war, and the justification for employing illegal torture against 'enemy combatants'; it is a clear case of the public discourse infiltrating the policy process. Moreover, it is an illustration of the extra powers that the state is attempting to invoke by declaring a condition of war: under American law, a state of war enhances the constitutional authority of the executive branch.

A second important sub-narrative is the notion that both the terrorist threat and the kind of war they have created are 'new' and unprecedented, and that consequently, a 'new paradigm' is required to successfully defeat them (one that may involve jettisoning 'old' restrictions). In large part, this is a reflexive discursive strategy designed to overcome the inherent contradiction involved in declaring a 'war' (and invoking national self-defence on the basis of international law) whilst simultaneously denying the applicability of the laws of war to captured fighters. The rhetorical solution is to declare that it is a 'new kind of war' fought not by recognized soldiers but by 'enemy combatants'. For example, Bush frequently stated that the war against terror was '*a different kind of war* that requires a *different type of approach* and *different type of mentality*,' he added, 'All of us in government are having to adjust our way of thinking about *the new war*.'[72] Similarly, John Ashcroft argued that the unprecedented threat posed by terrorism required '*new laws* against America's

---

70 Military Order of 13 November 2001, Detention, Treatment, and Trial of Certain Non-Citizens in the War Against Terrorism, in Danner, *Torture and Truth*, pp. 78-82.

71 Memorandum for Alberto R. Gonzales, Counsel to the President, from Jay S. Bybee, Re: Standards of Conduct for Interrogation under 18 U.S.C., 1 August 2002, in Danner, *Torture and Truth*, p. 155.

72 George W. Bush Jr, 'Press Conference', The East Room, Washington, DC, 11 October 11 2001, URL: http://usinfo.state.gov/topical/pol/terror/, accessed 29 August 2003.

enemies.'[73] In another major speech, Ashcroft stated that 'a *new* offensive against terrorism' was required, in which '*new* weapons', '*new* powers', and '*new* tools' of law enforcement would be utilized.[74] This language is deliberately employed to stress the unique circumstances of the war against terror; in such an unprecedented situation, it can easily be argued that the 'old' rules no longer apply.

This language and the thinking it engendered immediately found its way into the policy discussions on coercive interrogations. A memorandum to the cabinet on the treatment of prisoners in the Afghan theatre for example, stated: '[T]he war against terrorism ushers in *a new paradigm*, one in which groups with broad, international reach commit horrific acts against innocent civilians, sometimes with the direct support of states. Our Nation recognizes that *this new paradigm* – ushered in not by us, but by terrorists – requires *new thinking* in the law of war ...'[75] This language directly echoed the administration's public rhetoric and set the discursive foundation for the policies that were to follow, as the construction of a new kind of war and an unprecedented terrorist threat became one of the primary justifications for jettisoning decades of jurisprudence and practical lessons. In a memorandum for Alberto Gonzales, Assistant Attorney General Jay Bybee stated:

> As you have said, the war against terrorism is a *new kind of* war. It is not the traditional clash between nations adhering to the laws of war that formed the backdrop for GPW [Geneva Convention III on the Treatment of Prisoners of War]. The nature of *the new war* places a high premium on other factors, such as the ability to quickly obtain information from captured terrorists and their sponsors in order to avoid further atrocities against American civilians, and the need to try terrorists for war crimes such as wantonly killing civilians. In my judgment, this *new paradigm* renders obsolete Geneva's strict limitations on questioning of enemy prisoners and renders quaint some of its provisions...[76]

Similarly, in the report of a special working group on the legal and operational implications of detainee interrogations, it was suggested that:

> Due to *the unique nature of the war* on terrorism in which the enemy covertly attacks innocent civilian populations without warning, and further due to the critical nature of the information believed to be known by certain of the al-Qaida and Taliban detainees regarding future terrorist attacks, it may be appropriate for the appropriate approval authority to authorize as a military necessity the interrogation of such unlawful combatants

---

73 John Ashcroft, 'Testimony to House Committee on the Judiciary', 24 September 2001, URL: http://usinfo.state.gov/topical/pol/terror/, accessed 29 August 2003.

74 John Ashcroft, 'Prepared Remarks for the US Mayors Conference', 25 October 2001, URL: http://usinfo.state.gov/topical/pol/terror/, accessed 29 August 2003.

75 Memorandum for the Vice President, the Secretary of State, the Secretary of Defense, the Attorney General, Chief of Staff to the President, Director of Central Intelligence, Assistant to the President for National Security Affairs, and Chairman of the Joint Chiefs of Staff from President George W. Bush Jr., Humane Treatment of al Qaeda and Taliban Detainees, 7 February 2002, in Danner, *Torture and Truth*, pp. 105-7.

76 Memorandum for Alberto R. Gonzales, 1 August 2002, pp. 83-7.

in *a manner beyond that which may be applied to a prisoner of war who is subject to the protections of the Geneva Conventions.*[77]

In these texts, the logic and purpose of the reflexive language of the 'new' and 'different' war is clearly evident: because it is a fundamentally 'new' kind of conflict, a 'new paradigm' applies in which the previous limitations are rendered 'obsolete' and 'quaint', and interrogation may go beyond the protections of the Geneva Conventions. It is the logic of the discourse that drives the formation of the torture policy, rather than reasoned debate and consideration. This is why opposition to the policy was largely ignored and discounted.

A third sub-narrative that was crucial for structuring the torture policy was the discursive construction of a massive and ubiquitous terrorist threat. Although understandable in the aftermath of the devastating 11 September 2001 attacks, the public language regarding the ongoing danger of terrorism was nonetheless hyperbolic in the extreme. For example, according to administration officials, terrorism posed not just a threat of sudden violent death, but a 'threat to civilization', a 'threat to the very essence of what you do',[78] a 'threat to our way of life',[79] and a threat to 'the peace of the world'.[80] The Spokesman Coordinator for Counter-terrorism, Cofer Black, went even further: 'The threat of international terrorism *knows no boundaries*;'[81] in this formulation, terrorism was an infinite threat. In addition, administration officials suggested that the threat of terrorism was supremely catastrophic. Dick Cheney for example, stated: 'The attack on our country forced us to come to grips with the possibility that the next time terrorists strike, they may well … direct *chemical agents* or *diseases* at our population, or attempt to detonate *a nuclear weapon* in one of our cities.' He went on normalize the threat: '[*N*]*o rational person can doubt* that terrorists would use such weapons of mass murder the moment they are able to do so.'[82] Administration officials then went to great lengths to explain how the same terrorists (who are apparently eager to use weapons of mass destruction) are also highly sophisticated, cunning, and extremely dangerous: 'The highly coordinated

---

77 Working Group on Detainee Interrogations in the Global War on Terrorism: Assessment of Legal, Historical, Policy, and Operational Considerations, 4 April 2003, in Danner, *Torture and Truth*, p. 188.

78 Colin Powell, 'Remarks to the National Foreign Policy Conference', 26 October 2001, URL: http://usinfo.state.gov/topical/pol/terror/, accessed 29 August 2003.

79 George W. Bush Jr, 'Address to a Joint Session of Congress and the American People', 20 September 2001, URL: http://usinfo.state.gov/topical/pol/terror/, accessed 29 August 2003.

80 Bush Jr, 'State of the Union Address', 29 January 2002.

81 Cofer Black, Spokesman Coordinator for Counterterrorism, State Department, 'Press Conference for 2002 Annual Report 'Patterns of Global Terrorism', 30 April 2003, URL: http://usinfo.state.gov/topical/pol/terror/, accessed 29 August 2003.

82 Dick Cheney, Vice President, 'Remarks to the American Society of News Editors', The Fairmont Hotel, New Orleans, 9 April 2003, URL: http://usinfo.state.gov/topical/pol/terror/, accessed 29 August 2003.

attacks of 11 September make it clear that terrorism is the activity of *expertly organized, highly coordinated* and *well financed* organizations and networks.'[83] Moreover, officials argued that this was not a tiny group of dissidents; rather, 'there are *thousands of these terrorists* in more than 60 countries' and they 'hide in countries around the world to plot evil and destruction'.[84] Similarly, in a discursive construction that mirrored the mode of a popular spy novel, the president warned: 'Thousands of dangerous killers, schooled in the methods of murder, often supported by outlaw regimes, are now spread throughout the world *like ticking time bombs*, set to go off without warning.'[85]

The language of the national emergency engendered by the attacks, and the ever-present and potentially catastrophic danger posed by terrorists, entered the torture policy debate immediately. For example, the threat of further terrorist attacks is referred to in the Military Order of 13 November 2001, primarily as a means of establishing that a 'supreme emergency' is in effect:

> Individuals acting alone and in concert involved in international terrorism possess both the capability and the intention to undertake further terrorist attacks against the United States that, *if not detected and prevented, will cause mass deaths, mass injuries, and massive destruction of property*, and *may place at risk the continuity of the operations of the United States Government.* [...] Having fully considered *the magnitude of the potential deaths*, injuries, and property destruction that would result from potential acts of terrorism against the United States, and the probability that such acts will occur, I have determined that *an extraordinary emergency exists* for national defense purposes ...[86]

Similarly, in a direct echo of both the purported threat of weapons of mass destruction and the 'ticking time bomb' theory, and as a means of pre-emptively legitimizing the use of coercive interrogation, the Attorney General's office stated:

> ... al Qaeda has other *sleeper cells* within the United States that may be planning similar attacks. Indeed, al Qaeda plans apparently include efforts to develop and *deploy chemical, biological and nuclear weapons of mass destruction.* Under these circumstances, a detainee may possess information that could enable the United States to prevent *attacks that potentially could equal or surpass the September 11 attacks in their magnitude.* Clearly, any harm that might occur during an interrogation would pale to insignificance compared to the harm avoided by preventing such an attack, which could take *hundreds or thousands of lives.*[87]

Importantly, this text reveals how the logic of the constructed terrorist threat determines the moral calculations at the heart of the torture policy: based on the popular 'ticking

---

83 John Ashcroft, 'Testimony to House Committee on the Judiciary', 24 September 2001.

84 Bush Jr, 'Address to a Joint Session', 20 September 2001.

85 Bush Jr, 'State of the Union Address', 29 January 2002.

86 Military Order of 13 November 2001.

87 Memorandum for Alberto R. Gonzales, 1 August 2002, pp. 150-51.

bomb' myth in which a terrorist has been captured after planting a bomb in a secret location (like the plot of popular television programs such as *24* or *Spooks*), it is deemed morally expedient to torture the suspect in order to prevent an even greater evil from occurring. In fact, administration officials would have known that in thousands of cases of torture under similar presumptions, from Algeria to Israel, no bomb has ever been found.[88] The internal logic of the discourse however, means that such knowledge is discounted in favour of a predetermined course of action.

In sum, the documentary evidence suggests that the public discourse of the war on terrorism was more than just 'public diplomacy', 'public relations', or 'propaganda'. Rather, it functioned to inform and structure the policy-making process itself; in many ways, the official discourse dictated the logic of the torture policy as the public language found its way into the institutional deliberations and documents of the policy-making process. The discourses of counter-terrorism therefore, must be considered an essential element of the overall explanation for the abuses in Iraq and the wider war on terrorism.

The Language-Practice Connection

Apart from its direct influence on the policy-making process, it is also possible to demonstrate that the public discourse of the war on terror – in particular, narratives regarding the threat of terrorists, and the inherently evil, barbaric, and inhuman terrorist 'other' – are directly implicated in the actual practice of torture and prisoner abuse by individual soldiers. An examination of a specific instance of abuse, namely the torture revealed in the Abu Ghraib photographs, reveal the myriad ways in which the public discourse about terrorists was translated by individual soldiers into specific instances of abusive behaviour. For example, the extreme forms of shackling seen in the images of the initial Guantánamo Bay prisoners (in some cases, bound and shackled to gurneys, detainees were wheeled to interrogations; in others they were tightly shackled, blindfolded, and muzzled) were officially justified on the grounds that these were such dangerous individuals that they had to be restrained in this fashion for the safety of those guarding them. As General Richard E. Myers, chairman of the Joint Chiefs of Staff put it, they were such a threat because given half a chance, they 'would gnaw through hydraulic lines in the back of a C-17 to bring it down.'[89] Donald Rumsfeld said on a visit to Guantánamo that the prisoners there were 'among the most dangerous, best-trained, vicious killers on the face of the earth.'[90] In effect, the ubiquitous public narrative of highly trained, expertly organized, and fanatical super-terrorists was translated directly into abusive transportation and prisoner management practices. The fear generated by the discourse of the dangerous terrorist 'other' was thus discursively reflected in the prisoner control practices and the attitudes of the guards towards terrorist suspects.

---

88 Rose, *Guantánamo*, pp. 143-5.
89 Ibid., p. 2.
90 Quoted in ibid., p. 8.

Similarly, the public discourse by senior administration officials in which terrorists were frequently described as the *'faceless* enemies of human dignity'[91] was reflected in the institutional practice of putting hoods on prisoners or making them wear blackened goggles, masks, and ear covers during transit, thereby rendering them literally as well as figuratively 'faceless'. It is well-known that this practice increases the likelihood of abuse, as it effectively de-personalizes and de-individuates prisoners. Interestingly, in the Abu Ghraib abuse photographs the victims are hooded even though there is no practical need for them as they are restrained and already in custody. It seems reasonable to assume that in this instance, the hooding was both a deliberate means of de-humanization to facilitate the subsequent abuse and a subconscious attempt to confirm them as 'faceless' enemies. As the Milgram experiment and the real-world experience of counter-terrorism in Northern Ireland clearly demonstrated, interrogators find it far easier to inflict pain on their subjects when their facial expressions are obscured. In any case, these images reveal that the social and political construction of the 'faceless' enemy other was more than simply rhetoric or public relations; rather, it actually co-constituted the widespread abuses of the counter-terrorist campaign.

Public officials also frequently referred to terrorists as 'animals' and 'barbarians' who were outside the realm of civilized society. For example, America's ambassador to Japan stated that the 11 September 2001 attacks were 'an attack not just on the United States but on enlightened, *civilized societies* everywhere. It was a strike against those values that separate us from *animals* – compassion, tolerance, mercy.'[92] This language suggests that terrorists are both animals and barbarian savages. President Bush reaffirmed this formulation when he stated that: 'By their *cruelty*, the terrorists have chosen to live on *the hunted margin of mankind*. By their hatred, they have divorced themselves from *the values that define civilization* itself.'[93] John Ashcroft made a similar point: '[T]he attacks of September 11 drew *a bright line of demarcation between the civil and the savage*, and our nation will never be the same. On one side of this line are freedom's enemies, murderers of innocents in the name of *a barbarous cause*.'[94] In effect, this language placed terrorists outside of the civilized community, on the '*hunted* margins of mankind' and functioned to essentialize them

---

91 George W. Bush Jr, 'Remarks in Commencement Address To United States Coast Guard Academy', Nitchman Field, New London, Connecticut, 21 May 2003, URL: http://usinfo.state.gov/topical/pol/terror/, accessed 29 August 2003.

92 Howard H. Baker, Jr., US Ambassador, 'Japanese Observance Ceremony for Victims of Terrorism in the US', Tokyo, 23 September 2001, URL: http://usinfo.state.gov/topical/pol/terror/, accessed 29 August 2003.

93 Interestingly, President Bush made these comments close to the Great Wall of China, built specifically to keep out the barbarian hordes. George W. Bush Jr, 'Remarks by the President to the CEO Summit', Pudong Shangri-La Hotel, Shanghai, People's Republic of China, 20 October 2001, URL: http://usinfo.state.gov/topical/pol/terror/, accessed 29 August 2003.

94 John Ashcroft, 'Testimony to House Committee on the Judiciary', 24 September 2001.

as 'an evil and *inhuman* group of men'.[95] Effectively, it transformed them into sub-human savages and animals that needed to be hunted down and smoked out of their caves. Apparently, Ivan Frederick's favourite description of the prisoners was 'animals';[96] employing the same language as Bush and Rumsfeld, Frederick then had no compunction against treating them as 'animals'. The most visually powerful expression of this discursive rendering can be seen in the photograph of Lynndie England holding a prisoner on a leash. Disturbingly reminiscent of colonial era photographs of African slaves tied by the neck,[97] this image represents the ultimate realization of the discursive creation of the terrorist 'savage' or 'animal'. Similarly, the image of the 'savage' terrorist being confronted by a savaging dog is another discursive re-enactment of the public discourse: hunting dogs stalk the terrorist prey; the trained wild animal is employed to subdue the terrorist beast. In fact, military investigations reveal that the use of unmuzzled dogs to frighten and intimidate prisoners was a routine practice.[98]

A final narrative relevant here is the notion of the terrorist enemy as a kind of disease or sickness. Colin Powell for example, frequently referred to 'the *scourge* of terrorism'.[99] This medical metaphor associates terrorists with filth and decay. It was restated even more explicitly by Rumsfeld: 'We share the belief that terrorism is a *cancer* on the human condition.'[100] Bush in turn, spoke of the danger to the body politic posed by 'terrorist *parasites* who threaten their countries and our own.'[101] In these constructions, the terrorists are re-made as dangerous organisms that make their host ill; they hide interiorly, drawing on the lifeblood of their unsuspecting hosts and spreading poison. It is this image of the filthy, disease-ridden savage that perhaps subconsciously inspired the photograph of the prisoner smeared with what appears to be dirt or excrement.

---

95 Baker, 'Japanese Observance Ceremony for Victims of Terrorism in the US', 23 September 2001.

96 Quoted in Kevin Toolis, 'Torture: Simply the Spoils of Victory?', *Newstatesman* (10 May, 2004), p. 9.

97 Interestingly, the most famous photograph of the Abu Ghraib abuses shows a hooded man with his arms outstretched, standing on a box, and with what appears to be electrical wires attached to his fingers. In this image, and in another covert reference to American relations with African slaves (who were also viewed as sub-human), the hood and cloak appear to mirror the costumes worn by the Ku Klux Klan.

98 Hersh, *Chain of Command*, p. 36.

99 Colin L. Powell, 'Remarks by the Secretary of State to the National Foreign Policy Conference for Leaders of Nongovernmental Organizations (NGO)', 26 October 2001, Loy Henderson Conference Room, US Department of State, Washington, DC, URL: http://usinfo.state.gov/topical/pol/terror/, accessed 29 August 2003.

100 Donald H. Rumsfeld, Secretary of Defense, and Joint Chiefs of Staff, Gen. Richard Myers, 'Briefing on Enduring Freedom', The Pentagon, 7 October 2001, URL: http://usinfo.state.gov/topical/pol/terror/, accessed 29 August 2003.

101 Bush Jr, 'The State of the Union Address', 29 January 2002.

In the end, the virulent de-humanization of the terrorist 'other' leads directly to the literal attempt to de-individuate and de-personalize all terrorists – as well as the suspension of individual empathy and social inhibitions against wanton cruelty. The photographs of prisoners in huge piles of bodies, a mass of indistinguishable naked body parts and heads hooded to obscure individual faces, is the ultimate realization of this discourse. For a moment in time, the sub-human 'terrorists' are discursively remade as a squirming mass of parasites or cancerous cells; they cease to be individuals and their humanity dissolves. Disturbingly, the photographs of prisoners piled on top of each other also mirror the well-known images of piles of naked corpses in the concentration camps during World War II; the Holocaust too, was in part the result of a discourse that defined the enemy 'other' as inhuman 'animals' and 'parasites'.

In short, the abuse photographs represent more than simply the careless recordings of a few sadistic or psychologically ill individuals; they are in fact, the logical outcome of a powerful public and private discourse that systematically de-humanized, de-personalized, and demonized the enemy 'other'. Within the confines of this language, the resulting torture and abuse was more than unsurprising; it was highly predictable. It became completely normalized within the moral logic of the language and practice of the war on terror. Apart from the confidence with which the Abu Ghraib abusers conducted themselves – giving the thumbs up to the camera, posing naked bodies in the general corridor with other activities continuing on behind them – the abuse was so widely known and accepted that one of the pictures was reportedly used as a screen saver on a computer in the interrogation room.[102] Clearly, in the military practice of America's war on terror, the discourse had succeeded in erasing all sense of humanity and decency.

At the same time, part of the reason why the abuse scandal was so shocking to the American public, if only momentarily, was because the photographs themselves, while reflecting the binaries inherent to the discourse, also severely destabilized them. In these images, it was the American 'heroes' who looked like savage barbarians, animals, and evildoers, while the 'terrorists' looked like the innocent victims of American terror. This is why President Bush had to publicly assert that it was 'disgraceful conduct by a few American troops who dishonoured our country and disregarded our values';[103] the essential American identity had to be discursively rebuilt; the torture had to be re-made as 'unAmerican'. In fact, such actions are the most predictable consequence of discourses based on hate and fear.

A War of Terrorisms

The findings of this study have important theoretical and normative implications. From a theoretical perspective, they provide a level of validation for constructivist

---

102   Brody, 'The Road to Abu Ghraib', in Meerpol, *America's Disappeared*, p. 127.
103   Quoted in Brody, 'The Road to Abu Ghraib', in Meerpol, *America's Disappeared*, p. 113.

approaches to political analysis by revealing the ways in which public discourses infiltrate and structure the policy-making process. They demonstrate that political reality is in large part a social construct, manufactured through discursive practices and shared systems of meaning, and that language does not simply reflect reality, it also functions to co-constitute it. Critically, the findings of this chapter reaffirm that ideational and discursive analysis is crucial to our understanding of foreign policy and political practice.[104] More specifically, they demonstrate the ways in which projects of political violence are social and political constructions that emerge out of deliberative action by political entrepreneurs. Disturbingly, the process by which the ordinary 21-year-old Lynndie England was transformed into a torturer was the same process by which ordinary Serbs, Croats, and Hutus were induced to slaughter their former friends and neighbours.[105] Each instance involved the construction of a totalizing discourse of fear and threat, evil enemies, and justified counter-violence. Ontologically, this suggests that large-scale violence is never inevitable or causally predetermined by social structures or human nature; rather, it results from determined and carefully calculated political agency. The good news is that if political violence is constructed by human action, it can also be deconstructed by human action. In this case, ending the abuses of the war on terror requires speaking, thinking, and acting in alternative ways towards the enemy 'other'. As the Northern Ireland experience suggests (among others), it is only when politicians (and the media) abandon the language of demonization and de-humanization of the 'other' that the search for political solutions can begin to yield fruit.

Even more importantly, from a normative perspective this study confirms that the discourse of the war on terror has succeeded in undermining accepted democratic norms and social values of tolerance and non-violence. The language and practice of torture proves that the war on terror long ago graduated to a classic dirty war; no longer a war *against* terrorism, it is now a war *of* terrorisms – American and Coalition terrorism, and insurgent terrorism. In part, this is because at the most fundamental level, the construction of a large-scale project of political violence such as a 'war' on terrorism, entails the destruction of the moral consensus and the collapse of the moral community – and its replacement with discourses of victim-hood, hatred of the 'other', fear, and counter-violence. Once a society embraces these collective narratives, once it venerates its grievances and truly hates and fears an enemy 'other', public and political morality is quickly lost in the maze of national security expediencies. There is no starker illustration of the moral vacuity engendered by the public language of the war on terror than the policy and practice of torture. In some ways, this is a contemporary recalculation of the moral mathematics of Hiroshima, where '9-11' (the new and reconstituted 'Ground Zero') stands in for Pearl Harbor. According to this logic, if the torture/nuclear incineration of thousands of inhuman

---

104   See Alexander Wendt, 'Anarchy is What States Make of it', *International Organization*, 46, 1992, pp. 391-425; David Campbell, *Writing Security: United States Foreign Policy and the Politics of Identity*, Revised edition, (Manchester, 1998).

105   Jackson, *Writing the War on Terrorism*, pp. 180-2.

terrorists/treacherous Japanese will save American lives by preventing another 9-11/ Pearl Harbor, then it is morally acceptable. As Slavenka Drakulic expressed it, 'once the concept of "otherness" takes root, the unimaginable becomes possible.'[106] The once unimaginable has in fact, been realized in the policy and practice of torture and the accompanying public acquiescence to the abandonment of long held political and civil rights.

The simple reason for this tacit complicity is that these kinds of all encompassing and smothering discourses destabilize the moral community and replace non-violent political interaction with suspicion, fear, hatred, chauvinism and an impulse to violently defend the 'imagined community'. In addition, they automatically foreclose certain kinds of thought, simply because the language with which to frame doubts or question official justifications no longer exists or is inaccessible. While some individuals may initially feel unease at pictures of abused and humiliated 'terrorist' suspects at Camp Delta, of tortured Iraqi prisoners or dead Afghan civilians, they have no language or frame of reference in which to articulate those doubts. As time goes by, and when the discourse has been effectively absorbed by society, they may jettison such feelings altogether and consider the abusive treatment of suspects or the 'collateral damage' from bombing campaigns to be both justified and morally acceptable.

For this reason, as both responsible citizens and academics that live within a relatively open society, the only ethically responsible course left open is to vigorously oppose the language and practice of the war on terror. Apart from being morally corrosive and damaging to democratic values, the war on terror is also proving to be counter-productive and strategically ineffective;[107] in particular, after three years, the torture policy has failed to produce useful intelligence but succeeded in kindling greater anti-Americanism.[108] Alternative counter-terrorist approaches which do not rely on the de-humanization and demonization of an enemy 'other' are available and have a more reliable track record.

---

106   Quoted in Elizabeth Neuffer, *The Key to My Neighbor's House: Seeking Justice in Bosnia and Rwanda*, (London, 2001).

107   See among many others, Paul Rogers, 'International Security Monthly Briefing', The Oxford Research Group, URL: http://www.oxfordresearchgroup.org.uk/home.htm.

108   This is the considered conclusion of Rose, *Guantánamo*, and Hersh, *Chain of Command*, and many of the interrogation experts they interviewed. See also Watt, 'Torture, "Stress and Duress"', in Meerpol, *America's Disappeared*, p.105, and Brody, 'The Road to Abu Ghraib', in *America's Disappeared*, p. 116.

PART IV
War Crimes and Human Rights

Reconstructing War: The Politics of Conflict and Barbarity via the Yugoslav Tribunal

Tim Montgomery

'Massacres have become vital.' – Michel Foucault[1]

Understanding barbarity in warfare, is not so much a question of whether war is now more violent, and if so why. Although these are of course relevant questions, I would suggest that we need to take a step back to reflect upon how ideas of warfare are constructed. The legitimization process surrounding warfare, which such construction entails, serves many purposes. Any barbarity is concealed – what 'our' armed forces do is by default legitimate and there is no need to query their actions; any action that seems awry is the fault of a few 'rogue' soldiers. The legitimization process, more fundamentally, also negates the querying itself. War is said to follow a strict relationship of combatant to combatant, in a distinct period and locale of combat. Human rights are likewise said not to apply in war and any controls in wartime are self-regulatory. As such warfare is determined – to understand it is to understand its own terms.

Violence, and particularly the use of armed force by states, is a matter of normalization and legitimacy. It is essentially political. Its presentation, by contrast, is of war being merely an altered situation. The nation has become gripped by 'an emergency' and such a crisis has required that the 'normal' rules of society are suspended. Furthermore, as long as the objectives of the war remain 'military', that loss of life is merely 'necessary' and that suffering is not 'excessive', war may function as a credible activity.[2] Even with normality suspended, the legitimacy of the warfare, in effect, makes it normal. Regardless, warfare is anyway deemed of a finite occurrence, with clear sides, declarations and cessations, and purpose. Hence warfare, carried out in a legitimate fashion, is said to be both coherent and concise. Functionally and intellectually it is a neutral operation outside of politics.

---

1   M. Foucault, (Trans. R. Hurley), *The Will to Knowledge: The History of Sexuality Vol. I* (London, 1990), p. 137.

2   A. Roberts and R. Guelff, *Documents on the Laws of War* (3rd ed., Oxford, 2002), p. 513.

Such presentations in fact provide the key to an increased understanding of barbarization of warfare. 'Necessity', 'excessiveness' and 'validity' of target, as the central tenets of the legitimate war, are clearly subjective in definition. The state of crisis, likewise, suggests a limited period outside of control that requires stringent activity to return society to safety. Yet what is more important is the process involved in crisis. It is a cover for extraneous activity, in which, significantly, those who suspend, or rather supplant normality are also those who actively engage in the abnormality. Legitimization is therefore a process undertaken with political purpose. How we understand warfare contributes to its construction as a political activity. This process of construction, and the changing implications for an understanding of warfare, are the concerns of this chapter.

To help expose the constructions of warfare, two stark juxtapositions are made. First, a particular situation of warfare – the former Yugoslavia – that has been deemed a threat to international peace and security and assessed by an international court is used as a source both for analyzing warfare and for judging its functioning. The discussion draws on an analysis of the Judgements and, where informative, Indictments of the International Tribunal for the Former Yugoslavia, covering an eight-year period from its inception.[3] Second, refugees are added in to provide a particular manifestation of warfare. The United Nations Commission of Experts on the Former Yugoslavia concluded in its final report that 1.5-2 million refugees fled the conflict.[4] Beyond pure numbers, forced migrants also indicate the neutralization process entailed in warfare. The protection of those fleeing from warfare is a politicized process that specifically draws in ideas of human rights and legitimacy of action in warfare. In so doing this process places the nature of warfare in relief.

A framework to understanding the barbarity of war may therefore be developed. Brief treatments both of the idea of persecution of refugees in warfare and of the construction of war are offered. Subsequently, the analysis of the Tribunal suggests a necessary *re*construction of how warfare is understood, and the consequences of this. Such reflections have multiple implications, which are discussed at the end. There is a potential prioritization of human rights over sovereign policy. This thereby denotes how war is part of an ordering process, whereby present barbarity can be explained by a desire for policy effectiveness rather than shared morality. Here is a dichotomy of war. The use of mass violence is frequently challenged in its legitimacy. Yet, simultaneously, the use of armed force remains one of the few

---

3   The International Tribunal for the Prosecution of Persons Responsible for Serious Violations of International Humanitarian Law Committed in the Territory of the Former Yugoslavia Since 1991 (ICTY). Abbreviations of cases are used throughout. For full case details see http://www.un.org/icty/cases/jugemindex-e.htm. Full details of the establishment of the Tribunal are at www.un.org/icty. Indictments clearly have a specific juridical intent behind them, and are used as a supplement to Judgements, to provide background data on the conflict. Witness statements would have been a valuable resource, but are protected under Article 22 of the International Tribunal Statute.

4   Y. Beigbeder, *Judging War Criminals: The Problems of International Justice* (Basingstoke UK, 1999), p. 147.

activities open to states. Barbarity, therefore, becomes a marker for the retention of power, and an indication of how power seeks to survive. As barbarization becomes more understood, what is warfare's 'vitality' may also become its constraint.

The Politics of Definition

Refugees are a clear consequence both of armed conflict and of human rights violations. Therefore, forced migrants do certainly provide a useful practical example for an analysis of warfare. Likewise, the political linkage of human rights and armed conflict, in refugee terms 'persecution', is crucial. The concept of 'persecution' is central to the refugee definition under international and national law – 'the term "refugee" shall apply to any person who … owing to a well-founded fear of being persecuted … is outside the country of his nationality.'[5] Migrants from armed conflict therefore have political significance, in addition to the practical use of refugees to the present analysis. Controlling the rights of forced migrants is indeed deeply political. Reflections upon how persecution operates within the context of warfare are consequently most informative for understandings of barbarization.

Forced migrants indicate stark contrasts between standards of protection and the realities of politics. I would suggest that, in this globalizing age of shifting localities and structures of power, nation-states are reduced entities. As a site of power, the sole remaining forms of comprehensive control within the nation-state are over the waging of war and the movement of people. To simplify drastically, there is little else for states to do. This may form part of the explanation in the modern age for increases, numerically and in terms of suffering, in both migration and conflict. Regardless, the treatment of forced migrants, particularly at the time of large scale warfare, throws in to distinct relief how conflict, refugees and human rights are political.

The barbarity of current armed conflicts does cause mass flights of people. However, protection is limited. The UN's refugee agency has concluded that 'persons compelled to leave their country as a result of international or national armed conflicts' are not normally deemed to be refugees.[6] This is politics. To add in refugees from armed conflict would, at the time of one estimation, nearly double the number of forced migrants that the international community would need to recognize.[7] Accordingly,

---

5   United Nations, *The Convention Relating to the Status of Refugees* (1951), Article 1. A. (2). The Refugee Convention is presently the primary source of international refugee law, and thereby of the refugee definition. Available on the UNHCR website at http://www.unhcr. ch/cgi-bin/texis/vtx/protect/opendoc.pdf?tbl=PROTECTION&id=3b66c2aa10.

6   UNHCR, *Handbook on Procedures and Criteria for Determining Refugee Status under the 1951 Convention and the 1967 Protocol relating to the Status of Refugees*, UN Doc. HCR/IP/4/Eng/REV.1 1979 (reedited 1992) paragraph 164. See also G. Goodwin-Gill, *The Refugee in International Law* (2nd ed., Oxford UK, 1998), p. 139 and P. Kourula, *Broadening the Edges: Refugee Definition and International Protection Revisited* (The Hague, 1997), p. 110.

7   Garry estimated in 1998 that there were 13,200,000 recognized refugees in the world, but an estimated 22,729,000 if account were also taken of those fleeing armed conflict – H.R.

politics suggests that the identification of refugee status, particularly with regards to the other primary state activity – warfare – must be controlled.[8]

There is historical precedent here. Contradictions of practice and protection have been played out in the definitional process of 'persecution'. Abuse of human rights, particularly in wartime, has continued throughout the last century. Yet the concurrent forced migration has threatened to challenge the same structure of statehood that is responsible both for the warfare, for the violations and for the protection. Hence, usage of the concept in the modern era has been determined by an assertion of control over migration and warfare. In early twentieth century migrant protection regimes, the emphasis was solely upon the loss of nationality. Treaties on Russian refugees, Armenian refugees and people of Turkish, Assyrian, Assyro-Chaldean, Syrian and Kurdish origin made no reference to persecution; nor did the Convention Relating to the International Status of Refugees and treaties on forced migrants from Germany and the Saar.[9] By addressing only nationality, the intent was to stabilize the movement of *subjects* between Empires, especially during colonial wartime, and thereby stabilize the integrity of those states.

The Constitution of the International Refugee Organization, of 1946, did mark a first change.[10] More explicit in its protection for displaced persons and refugees, forced to flee because they were 'victims' of a regime, there is also the first reference to deportation for specific discriminatory reasons and to the idea of persecution. This is of course the period of mass regimes, severe turbulence and brutality, greater awareness of regionalization, if not globalization (whereby self-managed flight to another part of the world would be conceivable) and, arguably, an invigorated idea of human rights. Yet protection to refugees, comparable to the need within such mass wars, did not emerge. For example, although refugee law was indeed now regime-focussed (as opposed to loss of nationality-focussed), the home government of the forced migrant was given the right to comment on the case being presented.[11]

Politics has continued to dominate in the modern age. Priorities from 1951 to today have been to minimize the numbers of migrants to whom protection is given.[12]

Garry, 'The Right to Compensation and Refugee Flows: A "Preventative Mechanism" in International Law?', *IJRL* 10 (1/2) 1998, p. 97.

8   For a more extensive discussion, see P. Tuitt, *False Images: Law's Construction of the Refugee* (London, 1996), p. 15 and generally.

9   Treaties of 5 July 1922, 31 May 1924 (to which the Kurds were later added), 30 June 1928, 28 October 1933 and 10 February 1938 respectively. All Treaty references in A. Grahl-Madsen, *The Status of Refugees in International Law* (Vol. 1, Leiden, The Netherlands, 1966), pp. 122-131 and p. 132. He does refer, at p. 132, to refugees from the Nazi regime being persecuted. However, the treaty text only refers to loss of nationality.

10   Outlined in Grahl-Madsen, *The Status of Refugees*, pp. 134-5.

11   IRO Constitution Annex 1, Part 1, section c paragraph 1 (b) – Grahl-Madsen, *The Status of Refugees*, p. 136.

12   Other, that is, than when accepting refugees is politically expedient. Hungarian exiles were (relatively) enthusiastically accepted by Western European States, arriving as they did from the Cold War 'enemy'. Yet many such migrants had actually only fled the immediate

The Refugee Convention itself, whilst emanating out of a human rights declaration that had no temporal limitation, retained for a decade and a half its cut-off point for protection as flight prior to 31 December 1950.[13] It is suggested that the original drafters of the Convention, recognizing that new techniques of persecution were continuously invented, and perhaps mirroring what levels of barbarity had recently been experienced within 1930s and 1940s Europe, had deliberately tried to leave the definition open.[14] However, the international community has continued to contest the pervasiveness of what acts constitute persecution, particularly with regard to continuing armed conflict. There is no agreed definition.[15]

This is indeed reflected in supplemental arrangements. Extensive protection is given to the internally displaced (IDPs). Yet, it may be maintained, this reflects the 'more desirable' non-passing by these migrants of international borders, the internalization of costs within states and the proportionate distance from the West of such IDPs. Regional arrangements are also curtailed. For political reasons, situations of violent conflict are clearly separated from the persecution definition in the African and Latin American protection systems.[16] An attempt at harmonizing the EU's position reiterated an unwillingness to provide anything other than a restrictive definition of persecution. The primary focus instead is placed upon retaining competence to determine who shall be granted refugee status and upon what alternative forms of protection may be offered.[17] Hence, the consequences of warfare are ultimately concealed behind the perceived political necessities both of controlling people and of conflict.

The Construction of Warfare

Much may therefore be revealed by reflecting upon the refugee experience. Alongside this, a look at the construction of conflict itself allows further exploration of how barbarity operates and how it too may be concealed. Warfare is based on three core elements, one organizational, two conceptual. First, warfare is merely

---

security situation, fully intending to return – G. Coles, 'Approaching the Refugee Problem Today', in G. Loescher and L. Monahan (eds) *Refugees and International Relations* (Oxford, 1990), p. 376.

13 Article 1 A (2). The temporal limitation was removed by Article 1 (2) of the 1967 *Protocol Relating to the Status of Refugees*, for those State parties to that Protocol.

14 Grahl-Madsen, *The Status of Refugees*, p. 193. See also Kourula, *Broadening the Edges*, p. 91.

15 *Handbook on Refugee Status*, paragraphs 51-3.

16 Article 1 (2), *Convention on the Specific Aspects of Refugee Problems in Africa* (1969). Conclusion and Recommendation III (3), *Cartagena Declaration on Refugees* (1984). See Goodwin-Gill, *The Refugee in International Law*, Annexes.

17 European Union, *Joint Position on the Harmonised Application of the Term 'Refugee' in Article 1 of the Geneva Convention of 28 July 1951 Relating to the Status of Refugees*, EU Doc 96/196/JHA 4 March 1996, Article 4. On restrictiveness more generally in Western Europe, see Grahl-Madsen, *The Status of Refugees*, pp. 87-91. On alternative systems, see Kourula, *Broadening the Edges*, p. 64.

regulated. International laws of war, together with national Manuals of Military Law and Operational Handbooks, provide frameworks specifically for conducting warfare.[18] Read inversely, these indicate what certain key principles are inherently implied by states for themselves. Though not unlimited, it is permissible to injure the enemy into submission. Belligerents may act militarily (as long as those actions are 'proportional'), and act on the basis of military advantage. If targeting excludes protected persons, the enemy may be targeted.[19] Hence, war may be operated freely, framed with just certain self-defined constraints. As one Manual comments, the 'law of armed conflict is not intended to impede the waging of hostilities.'[20]

Second, war has a specific threefold construction. War is seen as explicitly *finite*, with discernable periods of conflict. This is important for judging whether it is war itself that is causing suffering. It has also enabled governments to assert, in the case of refugees, that change has occurred in their circumstances (the war is 'over') and they may be returned.[21] War is likewise constructed as about 'populations' (combatants, *hors de combat*, civilians). It is thereby perceived to be *generalized* and non-discriminatory, not targeting individuals for persecution. Finally war has traditionally been, by and large, *self*-regulatory. The terms for the practising and regulation of war are set by states and, with the exception of certain formal and informal approaches from third parties, the militarism itself is monitored by belligerent states.[22] It is for them to determine 'proportionality' and 'non-excessiveness'. Hence, control of warfare is an amalgamation of policy, practice and values, set in a context of legal interpretation.

Third, it is debated as to whether human rights are a framework separate from warfare and the laws of war. It is asserted, for example, that application and applicability of each make them different, as do levels of burdens of proof.[23] More fundamentally, human rights are deemed to be relevant to periods of peace, and laws of war at other times. Indeed, war is the ultimate 'emergency' in which human rights are said to be suspended. The International Covenant on Civil and Political Rights of 1966 allows for derogation from its content in a 'time of public emergency which threatens the life of the nation.'[24] However, this distinction between 'peace'

---

18 Roberts and Guelff, *Documents on the Laws of War*, pp. 4-14.

19 See Roberts and Guelff, *Documents on the Laws of War*, pp. 9-10.

20 US Dept. of Navy, *Commander's Handbook on the Law of Naval Operations*, as quoted by Roberts and Guelff, *Documents on the Laws of War*, p. 13.

21 *EU Joint Position*, paragraph 3. No consistent definition of the end of conflict is given, although a cease-fire might generally be used.

22 Formal approaches include the work of International Committee of the Red Cross, International Fact-Finding Commissions and, after the conflict, the International Court of Justice.

23 F. Maurice and J. de Courten, 'ICRC Activities for Refugees and Displaced Civilians', *International Review of the Red Cross*, 280, p. 10 and J. Meurant, 'Humanitarian Law and Human Rights Law: Alike yet distinct', *International Review of the Red Cross*, 293, p. 91.

24 Art. 4 1. Of course this is open to interpretation and such derogations are not absolute – see Art. 4 2.

and 'war' seems artificial and, thereby, is informative about how barbarization is constructed.

What occurs during war could not be validated under human rights of 'peacetime'. As has been commented, a 'war fought in compliance with the standards and rules of the laws of war permits massive intentional killing or wounding and massive other destruction that, absent a war, would violate the most fundamental human rights norms.'[25] Yet this is precisely the point. What is occurring is a de-prioritization of need and relative inattention to the victim.[26] The mass nature of suffering, inflicted in and out of war, should require attention to be on the victim. However the protection to those fleeing warfare is of a lower standard than for others. Ultimately then, I would suggest, human rights and protection are made secondary to definitional requirements flowing out of the practise of warfare. War has a specific construction. Therefore, to understand barbarity it is necessary to reflect upon how warfare operates and what political action it might be serving.

War and Barbarity: A View from the Tribunal

The Judgements and Indictments of the International Tribunal for the Former Yugoslavia are informative with regards to understanding the construction of warfare, and thereby also its barbarity. Four key areas are suggested. Analysis is given by the Tribunal of how acts by militaries are persecutory, and how their impact on victims cannot (or should not) be normalized. The supposed succinctness of warfare, one of its central constructions, is negated. Similarly, by exploring how armed conflict is itself persecutory, the barriers between what may be permissible due to crisis, and what is deliberately contrived, are broken down. Subsequently the necessity of inclusion of human rights, in the structuring of armed conflict, is expanded. As the application of human rights suggests a constraining of freedom to act military, with an inversion of rights over sovereignty, this creates a distinct dynamic of the modern age – barbarization may, despite such rights, become an increasing feature of warfare, as the desire to continue to fight wars outweighs the protection of those rights.

1. Persecutory Acts in Warfare

The barbarity of abuse of human rights is well-documented elsewhere. However, insights are given by the Tribunal into how persecution functions. This indicates thereby what it might mean to be 'persecuted'. Twelve particular characteristics of persecutory acts are evident, and are discussed briefly in terms of the acts and the victims.

---

25 H.J. Steiner and P. Alston (eds), *International Human Rights in Context: Law, Politics, Morals* (2nd ed. Oxford, UK, 2000), p. 68.

26 J. Dugand, 'Bridging the Gap Between Human Rights and Humanitarian Law: The Punishment of Offenders', *International Review of the Red Cross*, 324, pp. 445-6.

a) Act-focussed

i. Being a persecutory act   It is interesting that 'persecution' remains comprehensive and non-specified. International humanitarian law does contain a specific crime of 'Persecution', covered by the Tribunal's Statute at Article 5(h). Yet this is given a loose definitional framework, covering bodily and mental harm and infringements upon individual freedom.[27] The concept of 'persecution' more broadly is certainly treated openly by the Tribunal. The Prosecution in the opening case before the Tribunal successfully argued that persecution could be any inhumane act.[28] The suffering does not even need to be specific to a person or to type of act; the mere infringement of human rights in a discriminatory fashion is enough.[29] Persecution in warfare is therefore an act based on discriminatory intent for an arbitrary reason.[30] It covers, *inter alia*, killing, torture, physical violence, rape, sexual assault, humiliation, degradation, destruction of homes, religious and cultural sites and businesses, deportation and forcible transfer, and denial of fundamental rights.[31] It is, furthermore, an overarching crime against humanity, within which other crimes may be included and may, finally, indicate the *mens rea* for genocide and be a stage towards an act becoming genocidal.[32]

ii. Forced movement   According to the Tribunal, the large-scale creation of refugees can constitute persecution, even potentially amounting to genocide.[33] It may be implied, such as part of a population being killed to 'ensure the remainder would not want to return',[34] or linked to other actions, for example whereby it becomes likely that detainees will be deported.[35] The potential criminality of refugee-creation is striking. Both in the finding that breaches of international humanitarian law had taken place, and in the use by the Security Council of UN Charter Chapter VII powers to establish the Tribunal, the mass creation of refugees ('ethnic cleansing') is deemed a threat to international peace and security.[36]

iii. Economic persecution   Economic attacks are deemed by the Tribunal to be types of persecution. These can include termination of employment, the undermining

---

27 *Kordic* et al TCIII Judgement point 200.

28 As long as it was carried out with discriminatory intent against the civilian population on the specified grounds – *Tadic* Opinion and Judgement points 699,700 and 705.

29 *Tadic* Opinion and Judgement, points 697 and 715.

30 *Nikolic*, Review of Indictment IT-94-2-R61, T.Ch.1 part II. A. 5. The specific discriminatory grounds involved are what distinguishes it from genocide and most cases of ethnic cleansing – *Sikirica* Trial Chamber III Judgement on Defence Motion to Acquit Point 89.

31 IT-97-24-PT Fourth Amended Indictment point 54 (1)-(5).

32 *Sikirica* et al TCIII Judgement on Defence Motion to Acquit points 37 and 47; *Todorovic* Sentencing Judgement point 32 referring to Trial Chamber in *Blaskic*.

33 *Nikolic* Review of Indictment IT-94-2-R61, T.Ch.1, Part IV B. See also *Tadic* et al IT-94-1 TC I para 18 (i).

34 *Kovacevic* IT-97-24-I Amended Indictment point 31.

35 *Tadic* et al IT-94-1 TC II Indictment (2nd amendment) Count 1. Point 4.4.

36 *Tadic* et al IT-94-1 TC II para 21.

of livelihoods, attacks on specific professions, and the appropriation of property. Widespread extortion can form a part of a general process of intimidation, and threatening the means for working, such as capacity to work and the distribution of land, is persecution.[37] The Tribunal has held that, as long as the source of suffering is not a generalized economic process, then persecution can have an economic character.[38] Therefore, migrants, normally denied asylum as 'economic', could show persecution.

iv. Hindrance of general life   More generalized impediments to daily life are directly persecutory. These include discouraging the return of migrants, attacks on town halls, banks, post offices, telephone exchanges and courthouses, and limits on the use of a language. Similarly, the prohibition of the practising of worship, attacks on cultural heritage (such as buildings and objects used for religion, art, science, charitable purposes, historic monuments, hospitals and other medical units and education), and the banning of children from entering school are all crimes that occur in war. Finally, blockages to employment, travel curfews, delays in the issuing of documentation, the theft of cars, together with restrictions on access to petrol, the insufficient provision of health care and arbitrary and harsh judicial judgements are all persecution.[39] Such comprehensive treatments of armed conflict are important for breaking down normalizations of warfare.

v. Mental persecution   Not just physical attacks, but both the psychological trauma of a physical act, and the infliction of mental harm are persecution.[40]

vi. Indirect persecution   Although not necessarily facing direct abuse, a person can be barbarized by a lack of realistic choice. For example, witnesses described having to relinquish goods and to undertake not to come back to an area, in return for freedom from detention.[41] Furthermore, those who suffered the witnessing of crimes were considered to have been persecuted. The *Sikirica* case detailed how a wife and child were forced to watch an attack on their husband/father. Likewise, in that case, the holding of people in a room in Keratem camp, next door to the room used to torture and murder, was deemed persecution of the bystanders, as well

---

37 *Stakic* IT-97-24-PT Fourth Amended Indictment point 17d; *Karadzic* IT-95-5 Indictment para 40; *Karadzic* IT-95-5 Indictment para 40; *Jelisic* et al IT-95-10 Indictment point 31; *Tadic* et al IT-94-1 TC II Indictment (2nd amendment) 4.1; *Nikolic* IT-94-2 Indictment, point 21.1; *Tadic* et al IT-94-1 TC II Indictment (2nd amendment) Count 1, point 4.5; *Kordic* et al TCIII Judgement point 243; *Tadic* Opinion and Judgement point 136; *Sikirica* et al IT-95-8 Indictment point 33 and *Sikirica* Sentencing Judgement point 89; *Kordic* Trial Chamber III point 210.

38 *Tadic* Opinion and Judgement Point 707; *Jelisic* TC I Judgement Point 48.

39 *Karadzic* IT-95-5 Indictment para 29; *Tadic* Opinion and Judgement point 12; *Tadic* Opinion and Judgement 703; *Kordic* et al TCIII Judgement points 359 and 362; *Tadic* Opinion and Judgement points 149-150; *Simic* IT-95-9; *Kupreskic* et al TCII Judgement point 89; *Meakic* et al IT-95-4 Indictment para 28.1; *Tadic* Opinion and Judgement point 709; *Tadic* Opinion and Judgement point 703.

40 *Sikirica* et al IT-95-8 Indictment; Witness FWS-87 became suicidal after numerous assaults – *Jankovic* et al IT-96-23 Amended Indictment point 7.14.

41 *Tadic* Opinion and Judgement point 457.

as of the direct victim.[42] Similarly, even if complicit in, or forced to carry out, a barbarity themselves, it was concluded that the perpetrator was a victim of indirect persecution. One community was made complicit, by being informed of an imminent attack against their neighbours so that they had a chance to escape.[43] Examples of the actual commitment of crimes included Witness 'G' biting off a man's testicle, and individual detainees being made leaders of detention rooms or being required to beat other detainees.[44]

vii. Producing a climate of fear   The culture in which war is fought can be persecution. Such 'atmosphere[s] of terror'[45] might include prejudicial propaganda, the language that is publicly used, and control of the media.[46] In the generalized incitement and germination of criminal sentiment within the minds of the population, it is unclear whether the population must enact the suggestions for such propaganda to be deemed by the Tribunal as persecution. Yet it does seem enough that, whilst others are carrying out criminal acts, the population at large believes that these acts are both acceptable and necessary.[47] The fear is of living within a context of widespread human rights abuses.

b) Victim-focussed
i. Gender persecution   According to the Tribunal, women are persecuted because of their gender, regardless of whether there was a broader intent behind the acts. Although generalized fear was the purpose behind rape in Trnopolje Camp and the prohibition of non-Serb women giving birth in local hospitals was a feature of 'ethnic cleansing', the persecution was deemed specific to the women.[48] Certain detention camps, for instance the so-called 'Karaman's House', were established as brothels and slave labour points specifically containing women.[49] Although there was generalized detention in these sites, they had a gender-specific purpose over and above general intimidation. Sexual assault could be the *exclusive* form of abuse, rather than an associated or secondary act. Trnopolje Camp, for instance, was

---

42 *Sikirica* et al IT-95-8 Indictment.

43 *Kupreskic* et al TCII Judgement points 333-4.

44 *Tadic* et al IT-94-3 Indictment para 5.1, and *Sikirica* Sentencing Judgement point 66 and 98 respectively.

45 *Jelisic* et al IT-95-10 Indictment point 41. The acts which go to creating this atmosphere need have no specific purpose, other than fear, and are thereby different to other abuses (such as torture or biological experimentation) – *Kordic* et al TCIII Judgement point 243.

46 *Meakic* et al IT-95-4 Indictment para 27.1; *Stakic* IT-97-24-PT Fourth Amended Indictment point 17 b.

47 *Tadic* Opinion and Judgement point 708.

48 *Tadic* Opinion and Judgement point 175 and *Tadic* Opinion and Judgement point 147 respectively. Similar such acts were inflicted on men, as well as women, but this was not deemed to detract from the specific genderised purpose – *Tadic* et al IT-94-3 Indictment para 2.6.

49 *Jankovic* et al IT-96-23 Amended Indictment points 2.1, 7.5 and 8.3.

established as a deliberate sexual assault site.[50] The abuse could also be indirect, yet still be gender based. Women were abused to place pressure on men, such as where they were interrogated to reveal the whereabouts of opposition forces.[51] Ethno-political abuse had a specific gender-focussed tone, such as a rape victim being told that she would 'now give birth to Serb babies.'[52] Certain acts, such as rape and sexual assault, were also reconstituted in the Tribunal as torture.[53]

ii. Implied persecution the basis for persecution may be imputed, inferred or imposed.[54] Crimes against humanity were carried out on persons 'who were or were *believed to be* the Muslims or non-Serbs.'[55] People were attacked for their purported voting patterns. Jovo Radocaj was beaten to death because of a belief that he supported the Bosnian Muslim Party for Democratic Action. Jasmir Ramadanovic was beaten after being accused of being a 'Muslim extremist'. Adnan Kucalovic was beaten after being accused of having a brother fighting for the Muslim resistance.[56]

Imposed status likewise becomes itself persecution. For example, Serbs were required to undertake loyalty tests that would show their allegiance to the persecutory program.[57] Residents of Stupni Do village thought themselves to be part of Bosnia-Herzegovina, but the leaders of the Croatian Defence Council decided they were part of the Croatian Community of Herceg-Bosna and treated them accordingly.[58] Hence it is how the belligerent perceives the 'enemy' that can be barbaric.[59]

iii. Opposition forces The purpose of conflict, even with all constraints on excessiveness and necessity, is to gain the 'partial or complete submission of the enemy'.[60] This of course makes those in opposition vulnerable during armed conflict. More specifically, in an era mainly not of state-state conflict, or of civil war, but in

---

50 *Jankovic* et al IT-96-23 Amended Indictment points 6.4 and 6.5. On Trnopolje Camp, see *Kovacevic* IT-97-24-I Amended Indictment point 29. To this may be added where women were also forced to be sexual and domestic slaves – *Jankovic* et al IT-96-23 Amended Indictment point 8.6.

51 *Jankovic* et al IT-96-23 Amended Indictment point 5.3. See also *Jelisic* TCI Judgement point 39.

52 *Jankovic* et al IT-96-23 Amended Indictment points 7.9 and 7.17.

53 *Nikolic* IT-94-2-R61, T.Ch.1 Review of indictment Part IV A.

54 *Jelisic* TC I Judgement Point 71.

55 *Nikolic* IT-94-2-R61, T.Ch.1 Indictment point 23.1. Emphasis added.

56 *Sikirica* et al IT-95-8 Indictment points 25, 21 and 38 respectively. See also *Sikirica* Sentencing Judgement point 89 and *Nikolic* IT-94-2-R61, T.Ch.1 Indictment part II A. 2 (b). Clearly, being an 'extremist' is to be defined by someone else, as everyone would deem their position to be the starting point.

57 *Kovacevic* IT-97-24-I Amended Indictment point 39. Such persecution perhaps follows the traditional model of understanding of the concept, but on a localized scale; that is the need for freedom from authoritarianism.

58 *Rajic* IT-95-12 Indictment points1-2.

59 Emphasis added. *Sikirica* Trial Chamber III Judgement on Defense Motion to Acquit point 88 quoting Trial Chamber in *Jelisic* case para 70.

60 US Dept. of Navy's Commander's Handbook on the Law of Naval Operations, as quoted by Roberts and Guelff, *Documents on the Laws of War*, p. 10.

which oppositions may be routinely denounced as 'terrorist', the protection of such groups is particularly significant. Certain extensions to the protection regime are made by the Tribunal. Opposition forces generally are granted certain protection.[61] Certainly, under no circumstances may crimes against humanity be committed against someone because they are a member of a non-state group.[62]

Nor may supporters of opposition groups be persecuted. Although giving encouragement to a resistance force does engage a person in the conflict, it makes them neither a combatant nor permits their oppression.[63] Finally, political opinion may not be suppressed during wartime. Whilst counteracting 'activities hostile to the security of the State' is permitted under international humanitarian law, this only includes responding to 'espionage, sabotage and intelligence with the enemy Government or enemy nationals.' It specifically excludes suppressing 'a civilian's political attitude towards the State.'[64] Such judgements do provide insights into the so-called War on Terror, and whether the apparent state of emergency does allow suspension of such increasing standards of protection.

iv. 'Populations'   Recognition has emerged that conflict is not merely generalized violence, but can be specifically discriminatory. Whilst war is declared upon a nation, or on a population group, the Tribunal determined that victimization of an individual, through their membership of a collective, is no less persecution than direct abuse. The conceptual linkage between the crime of 'persecution on political, racial, religious or ethnic grounds' and that of 'institutionalized discrimination on racial, ethnic, or religious grounds involving the violation of fundamental human rights and freedoms and resulting in seriously disadvantaging a part of the population' is most significant.[65] Hence, neither the persecutory intent, nor the suffering and need for protection of the victim, are mere by-products of the conflict, but rather key factors in the nature of warfare.

v. Social groups   Indeed, how people are understood as a collective can be a source of their victimhood.[66] Although avoiding over-extravagance of claim,[67] the trend does seem to be towards an expansive definition of persecutable groups. These include, most comprehensively, political, religious, social, cultural and language groups.[68] Furthermore, it is the suffering involved, not the centrality of the practice that is crucial. Hence persecution may be by virtue of being 'prominent and highly

---

61 *Rajic* IT-95-12 Indictment.

62 *Tadic* Opinion and Judgement point 643 and *Jelisic* TCI Judgement point 54.

63 *Sikirica* TCIII Judgement on Defence Motion to Acquit point 38.

64 *Kordic* TCIII Judgement point 280.

65 *Tadic* Opinion and Judgement point 697. See also point 644.

66 This is particularly important for the international refugee definition, as one determining factor for persecution is that it is 'for reasons of ... membership of a particular social group', Refugee Convention Article 1 A (2).

67 Fishermen were attacked, but it would be hard to claim that fishermen are a 'particular social group' – *Jelisic* et al IT-95-10 Indictment point 31.

68 Here the Tribunal drew upon the International Law Commission's 1991 Report – *Tadic* Opinion and Judgement point 703.

educated', being 'political, economic, professional, academic and civil leaders' or 'intellectuals, professional and political leaders, and military aged males.'[69] Persecution is both an abuse of a group's ability to function and an attack on what it represents. What is being done here is juxtaposing daily lives, and the ability of people to live them, against the affects of warfare. The former is prioritized, rather than allowing warfare to be merely a dominant artifice.

2. Duration and Severity of Warfare

The apparent succinctness of war is one of its primary constructions. Declarations of war to cessation of combat and peace accords provide a concise and precise definition of warfare. Yet it is significant to reflect on whether this is an accurate construction of warfare. Indeed it is also of practical interest for the forced migrant – the duration and severity of the persecution are key dimensions of both the refugee definition itself (how 'well-founded' is the persecution) and the denial of protection to many forced migrants, particularly those fleeing war. International protection is generally seen as a temporary measure, as the cause of the flight, the conflict, is of finite time and impact.

International humanitarian law does of course already allow for protection outside of direct combat. It applies up to the 'general conclusion of peace'.[70] Nonetheless, a victim-focussed approach to conflict recognizes the long-term duration of impact of warfare. Generally, the Tribunal found that it was the cumulative nature of barbarity in warfare, rather than the individual acts, that was the greatest crime.[71] Duration of action was, furthermore, an aggravating factor in sentencing; it being held that the longer the actions lasted, the more opportunity the defendant had to stop them and to show remorse.[72] More specifically, the seizure of property, businesses and homes, for instance, would make return extremely difficult.[73] The effects of torture, sexual violence (particularly rape) and the installation of fear are of course long-lasting.[74] The

---

69 Quotes from *Stakic* IT-97-24-PT Fourth Amended Indictment, point 49 (2), *Simic* et al IT-95-9 Indictment point 5 a and *Kovacevic* IT-97-24-I Amended Indictment point 8 respectively. On the latter point, it could indeed be argued that war specifically persecutes males, in particular, from teens to late middle age, as they are deemed potentially recruitable and therefore a target. For the consequences of this, see the impact on the villages of Jaskici and Sivci in *Kovacevic* IT-97-24-I Amended Indictment point 35 and on the Brdo region in *Tadic* et al IT-94-3 Indictment para 11.1. Additionally see *Sikirica* et al TCIII Judgement on Defense Motion to Acquit point 80 and *Jelisic* TCI Judgement point 21.

70 *Tadic* Opinion and Judgement, point 633, quoting Appeals Chamber.

71 See, for example, *Sikirica* Sentencing Judgement generally and *Kordic* et al, TCIII Judgement point 205.

72 *Todorovic* Sentencing Judgement point 63.

73 *Tadic* et al IT-94-1 TC II Indictment (2nd amendment) 4.5 Count 1.

74 On torture, see reference to the statements of Hakija Elezovic point 288 *Tadic* Opinion and Judgement. On sexual violence, see *Tadic* et al IT-94-1 TC II, point 46, Vasif Gutic, prisoner assigned to medical unit at Trnopolje Camp, *Tadic* Opinion and Judgement point 175

impact of propaganda may be truly long-term.[75] Hence persecution within warfare may be said to be both ongoing and long-term. Importantly, what is largely perceived of as 'war' is in fact only one stage in a long-standing conflict of a group or personal nature.[76]

3. Persecution and Armed Conflict

The expansiveness of the Tribunal on how military action is specifically criminally barbarous is one of its most revelatory aspects. Insights are provided into how persecution operates within an armed conflict and, most notably, how armed conflict and military action should be contextualized as being within persecutory programs. Of course, a transition from prisoner of war camps to camps for the purpose of torture is an explicitly dramatic shift in terms of delegitimate military activity. Camps were 'deliberately operated in a manner designed to inflict upon the detainees conditions intended to bring about their physical destruction.'[77] However the Tribunal also re-conceptualizes what might be deemed 'regular' military activity. The instilling of fear, which is persecutory, is itself a military action.[78] The intent to 'kill, terrorize, or demoralize' a community, militarily, is a crime.[79] The infringement of human rights, even if part of military operations, amounts to a crime. Accordingly, there is an implied right, as a civilian, to be able to survive war; preventing this, for example by attacking aid convoys or providing insufficient protection, is illegal and persecutory.[80]

What is then both 'military', and within that persecutory, needs to be reconstructed. It is useful, at minimum, to introduce to our understandings of warfare the idea of 'wanton' violence, in which acts have no perceivable purpose or merely add to the creation of a culture of violence.[81] More formally, a military policy of dominance can be persecution, such as in the take-over of Prijedor, neither through a democratic process, nor as part of military 'spoils', but as a systematic long-term plan to seek dominance of a territory.[82] Even the fulfilment of other agendas within the wider context of conflict, where it can be shown that there is 'actual or constructive' knowledge of that wider context, do amount to crimes.[83]

---

and *Jankovic* et al IT-96-23 Amended Indictment point 6.5 and 7.7. On the effect of fear see Witness FWS-95, *Jankovic* et al IT-96-23 Amended Indictment, point 6.11.

75 Such as the speeches of Radoslav Brdanin, referred to in *Tadic* Opinion and Judgement point 89.

76 *Sikirica* Sentencing Judgement point 89.

77 *Kovacevic* IT-97-24-I Amended Indictment point 28. See also *Tadic* et al IT-94-3 Indictment para 2.2.

78 See, for instance, *Sikirica* et al IT-95-8 Indictment Part II.

79 *Karadzic* IT-95-5 Indictment para 26; *Kordic* et al IT-95-14 Indictment point 33.

80 *Kordic* et al TCIII Judgement point 705-6; *Kupreskic* et al TCII Judgement point 90.

81 See, for example, *Nikolic* – Review of indictment IT-94-2-R61, T.Ch.1 part II. A. 1 (a).

82 *Tadic* Opinion and Judgement point 138.

83 *Tadic* Opinion and Judgement points 656-9. Tadic's beatings of Sefik Sivac as revenge for being thrown out of the latter's café were a crime, *Tadic* Opinion and Judgement point 267.

Similarly, acts having seemingly a strictly military character may have an underlying persecutory motive – the military destruction of property preventing the return of populations to an area;[84] the shelling of markets to demonstrate military capability, to prevent other military actions and to be generally intimidatory;[85] the destruction of town halls, banks, post offices, telephone exchanges, courthouses as an affront to humanity and dignity;[86] the destruction of sites even if they are outside the general field of battle (such as attacks by Serb armed forces on sacred sites within Banja Luka, despite being in an area that was controlled by Serbs and thereby away from the war);[87] attempts to remove international forces or refugees.[88] Even the specific use and mobilization of the military infrastructure can be deemed criminal.[89]

Persecution can therefore be said to have taken place in the context of armed conflict. By finding these acts to be crimes, the victims' suffering is deemed no less grave than if they had been targeted outside war. Wartime does not legitimize the occurrence of acts which would not be allowed in peacetime.[90] Furthermore, it was early established that acts do not need to be 'in the heat of battle' to be criminal, but merely 'closely related to the hostilities'.[91] Armed conflict is therefore a source of crimes, as well as the context for the crimes and, as such, a reappraisal is required of traditional perceptions of the nature of war and its effects.

4. A Human Rights Focus

One of the most interesting insights from the Tribunal is that persecution in armed conflict is framed within an understanding of human rights.[92] This moral and legal framework requires a radical shift, to accommodate ideas of rights so 'fundamental' that they are 'of humanity'. The Tribunal holds that human rights must be implemented without discrimination, derogation (derived from their inherence to human existence) or divisibility.[93] Even more significantly, it also conceives of human rights, and thereby persecution, as a *process*. Protection against cruel treatment, and any

---

84 *Karadzic* IT-95-5 Indictment para 29.

85 *Kordic* et al TCIII Judgement Point 670.

86 *Tadic* Opinion and Judgement point 125.

87 *Karadzic* IT-95-5 Indictment para 37.

88 *Karadzic* et al IT-95-18 Indictment. *Kordic* et al TCIII Judgement point 8. *Simic* et al IT-95-9 Indictment and *Kovacevic* IT-97-24-I Amended Indictment point 1.

89 *Jelisic* TC I Judgement Point 53.

90 *Aleksovski* Judgement Point 37.

91 *Tadic* Opinion and Judgement points 632 and 573.

92 *Todorovic* TCI Sentencing judgement point 12 (g) and again at point 46; *Sikirica* et al IT-95-8 Indictment; *Tadic* Opinion and Judgement points 703 and 715.

93 See *Tadic* Opinion and Judgement points 707 and 715, *Sikirica* et al TCIII Judgement on Defence Motion to Acquit point 41 and *Tadic* Opinion and Judgement point 710 respectively.

subsequent prosecutions, are a 'means to an end'[94] – the 'end' is the reestablishment of humanity. This framework therefore requires that protection from persecution is an objective of the international community. More fundamentally, being part of an individual's dignity, human rights inherently supersede any policy or institutional privilege for war.

The Function of Barbarity and What May be Revealed

It may seem a tautology to say that politics is political. However it is worth stressing that policy actions do not operate in a bubble of neutral administration. The treatment of refugees and the functioning of war are specific political occurrences. This is an important recognition for an understanding of barbarization. How it is conceived – its specific construction – directly informs how its consequences are dealt with.

It might be argued that the conflict in former Yugoslavia was of such a heinous nature that it is not indicative of all war. However this would be to misconstrue the purpose of the Tribunal and its wider significance. The Tribunal was charged solely with prosecuting those responsible for 'serious violations of international humanitarian law.'[95] It does not however follow that less grave acts, with which it is not concerned, do not occur or are of no relevance. As the Tribunal itself recognizes, many acts, which are most serious but not grave, will not be tried.[96] Indeed, even the Tribunal asserts a role, in understanding the barbarity of warfare, outside of prosecutions alone. There is a desire to develop an idea of the context in which crimes take place. The Tribunal also performs a reconciliation function.[97] Above all, such deliberations contribute to our construction of 'humanity'. Juxtaposing what is deemed legitimate in society against any particular ordering process of political power is a deeply significant act. Studying the barbarous nature of warfare in general can only add to such understandings.

At minimum, what reflections, such as those of the Tribunal, do is to attempt to place human rights above warfare. I would suggest it is no coincidence that impetus has been given, by the occurrence of wars, to the development of the general human rights regime – for example, the discernible causal relationship between World War II and the UN Charter, the Geneva Conventions, the Universal Declaration of Human Rights and the Genocide Convention.[98] War is neither a normal nor risk-free activity, and as such a framework for protection is required. More importantly, war does not indicate an absence of politics, but an application of it. Therefore it

---

94 *Tadic* Opinion and Judgement point 723.

95 ICTY Statute, Article 1.

96 *Sikirica* et al TCIII Judgement on Defence Motion to Acquit point 139 and *Tadic* Opinion and Judgement point 572.

97 On its role in the promotion of peace, see part I 26 of Report of Secretary General under Security Council Resolution 808 UN Doc S/2504 3 May 1993, as referenced in Steiner and Alston, *International Human Rights in Context*, pp. 1147-8.

98 Steiner and Alston, *International Human Rights in Context*, p. 112.

is most significant that human rights, a structure of the *individual* rather than the *subject*, should be being applied in wartime. The Tribunal explicitly sees itself as dealing with crimes that are essentially *against humanity*. Hence, whilst an attack is on a specific person or group, these acts destroy heritage and culture that are common to, and inherently are shared with, the rest of humankind.[99] The Tribunal determined that its legitimacy comes essentially from an international consensus 'on what is demanded of human behaviour'.[100] Thereby it relies upon a perceivable ethical, normative behaviour system, one which is constituted from 'elementary considerations of humanity'.[101]

Of course, the 'human rights problem' remains. Those who primarily commit abuses – states – are also those who are charged with defining them, enabling them and of expressing 'international opinion' about them. They too are those that retain the legitimacy of use of armed force that may be being curtailed here. Nonetheless, the mere bringing in of human rights to warfare is significant.[102] It is important for forced migrants, for the refugee protection regime is derived from human rights standards.[103] Human rights are beginning, furthermore, to provide a structure to the practice of war and to the construction of permissibility of action in warfare. The laws of war are being added to and potentially even superseded.[104] Hence the dichotomy between 'laws of wartime' and those of 'peacetime', the latter being human rights, is being broken down.

It is in this context that decisions on withholding human rights protection in wartime are being made. For example, in a case before the UK High Court, for the families of six Iraqis killed by the UK military, the UK Government had argued that UK state military forces were not bound by human rights legislation in conflict situations. The justices found that the UK Human Rights Act does apply across all locations under UK jurisdiction, including prisoner of war detention facilities. Yet, fascinatingly, it does not apply to killings 'in the field', as military forces are operating 'outside the jurisdiction of the United Kingdom'.[105]

---

99   *Kordic* et al TCIII Judgement point 207. In this case, the specific reference is to attacks on religious sites, but the proposition can be made in all crimes against humanity.

100   *Tadic* et al TC II point 6.

101   *Tadic* Opinion and Judgement points 610 and 612.

102   To some extent the Tribunal side steps the 'human rights problem'. By being established under the United Nations Charter and implementing and developing international law, it deems itself to be a reflection of the will of the international community. Hence it is an international consensus 'on what is demanded of human behaviour' – *Tadic* et al TC II point 6.

103   The opening assertion of the Preamble to the 1951 Refugee Convention. See also Goodwin-Gill, *The Refugee in International Law*, p. 139.

104   T. Meron, 'International Criminalisation of Internal Atrocities', in *State Crime*, D.O. Friedrichs (ed.) (Vol. II, Ashgate, UK, 1998), p. 213.

105   Case No: Co/2242/2004, In The High Court Of Justice, Queen's Bench Division, Divisional Court, 14 December 2004, [2004] EWHC 2911, Between: The Queen – On The Application Of – Mazin Jumaa Gatteh Al Skeini And Others and The Secretary Of State For

On the specific level of warfare itself, how barbarity functions in warfare can increasingly be shown. Traditional perceptions of war as a discreet and finite occurrence are limited. The idea of 'generalized violence', that is not persecutory, presumes both clear commencement and termination points for conflict and the meeting of distinct 'sides' of armed forces. This is not the nature of modern war.[106] Conflict exists when armed force is used or when protracted armed violence is underway.[107] Nor should conflict be treated in isolation. It is sourced by and operates within a framework of instability, caused for example by decolonization, economic deprivation and resource scarcity. It is exacerbated by the availability and devastating potential of weaponry. Therefore, it seems appropriate to conceive of violence more broadly. Persecution does take place in conflict situations. The conflict can, of itself, be persecutory and provide a context for persecution. Persecution can also be part of a continuum; peace is a process, not the mere absence of war.

The Tribunal has also expounded upon the persecutory nature of certain acts, where they inflict harm or cause the suffering of indignity. The intent behind the acts could be random or purposeful. The targets may be specified or there may be being created a generalized context of persecution. The Tribunal also provides insights into the plight of particular victims. As well as more traditional ideas of 'protected persons', victims of persecution may be language groups, those in safe havens and other internationally designated persons, people in particular professions and people of certain ages and gender. Victimhood may be incidental, may reflect a persecution building upon socially normalized discrimination, or may be unique to the specific conflict.

How victimhood is operated is of wider significance. To contest the legitimization of targeting (including supposedly non-targeted 'collateral damage') is to contest the political subjectification of people. Broadly then, a specific barbarity of warfare is the effect it has on whole populations. This may be witnessed in a number of ways. Conscription, the use of reserves, or even an appeal to 'security moms'[108] all reflect how warfare stretches beyond the proclaimed battlefield. The role of torture, deportation and internment likewise indicate a broader pervasiveness of warfare on society. I would include in this situations in which it is attempted to place 'enemy combatants' in camps outside of all societal structures[109] or turning whole countries

---

Defence and The Redress Trust; available at http://www.redress.org/news/Judgment%20Al% 20Skeini%2014%20Dec%202004.pdf viewed on 16 June 2005.

106 A.C. Helton, 'Legal Dimensions of Responses to Complex Humanitarian Emergencies', *IJRL* 10 (3) July 1998, p. 538.

107 *Tadic* Decision on Appeal T.Ch.1 Part III B 1.

108 Many women in the US general election of 2004 were deemed by some commentators to have voted on the basis of which candidate would protect them most by fighting wars on 'terror' on their behalf – see for instance A. Swanson, 'Analysis: "Security" moms decided election', *Washington Times*, 2004, http://washingtontimes.com/upi-breaking/20041112-035043-8988r.htm, viewed on 16 June 2005.

109 'Enemy combatants' is a classification used by the US Government, which it has argued allows it to detain those persons without recourse to US or international law, but

into torture units for other states.[110] Finally, to fight wars in the name of the population, such as in taking 'the fight to the enemy [which] has made America safer',[111] is not just rhetoric, but is a form of complicity. Accordingly all 'subjects' become potential targets for retaliation.

In analyzing the barbarity of warfare, much may be being exposed. There is a use in maintaining a pretence of acceptability and legitimacy in warfare, for it may be concealing distinct changes in the nature of power in a globalized world. Globalization, in its many forms, challenges sovereign power. The purview of states is increasingly minimized to ever-decreasing regulation. The two remaining truly active roles, it may be argued, are, indeed, to wage war and to control the movement of people. That said, even here a number of challenges are faced.

Armed conflict is now no longer *de facto* solely between states, or even within states. Non-state agents take up arms of varying kinds and operate functionally, such as against economic symbols, rather than territorially. The international criminal legal system has, somewhat controversially, now recognized this, holding all persons, whether of non-state or state, criminally responsible and accountable for human rights.[112] Whether the barbarity of non-state groups is a response to that of states, or vice-versa, may be debated. Nonetheless it is worth looking beyond the normalization process for state armed force.

It seems possible that the retention of sovereign power over people and over warfare is being prioritized. Effectively undermining the International Criminal

---

merely 'in a manner' that is 'consistent with' the 'principles' of such laws (see for example US Department of Defense, 'Secretary Rumsfeld Remarks to Greater Miami Chamber of Commerce', 13 February 2004, http://www.defenselink.mil/transcripts/2004/tr20040213-0445.html viewed on 16 June 2005). The US Supreme Court has contested the Government's assertion that the location of the detention, Guantanamo Bay, is outside of US judicial responsibility – 542 U. S. ＿＿ (2004) 1, Opinion of the Court, Supreme Court Of The United States, Shafiq Rasul, et al., v. George W. Bush, President Of The United States, et al., Fawzi Khalid Abdullah Fahad Al Odah, et al., v. United States et al. [June 28, 2004], http://www.supremecourtus.gov/opinions/03pdf/03-334.pdf viewed on 16 June 2005.

110　The NGO Human Rights Watch recently reported on an 'increasing number of governments [that] have transferred, or proposed sending, [people] to countries where they know the suspects will be at risk of torture or ill-treatment. Recipient countries have included Egypt, Syria, Uzbekistan, and Yemen, where torture is a systemic human rights problem. Such transfers have also been effected or proposed to countries such as Algeria, Morocco, Russia, Tunisia, and Turkey' – Human Rights Watch, 'Still at Risk: Diplomatic Assurances No Safeguard Against Torture', 2005, http://hrw.org/reports/2005/eca0405/ viewed on 16 June 2005.

111　President G.W. Bush, 'State of the Union', 2 February 2005, http://www.whitehouse.gov/stateoftheunion/2005/index.html#4, viewed on 16 June 2005.

112　*Kordic* et al TCIII Judgement part III; *Tadic* Opinion and Judgement point 654; *Nikolic* IT-94-2 Indictment points 6.1-6.3. This is now transferred to the Statute of the International Criminal Court which, at Art. 7 2.(a), refers to acts 'pursuant to or in furtherance of a State *or organizational* policy' (emphasis added) and notes, at Art. 27 1., that the Statute applies 'equally to all persons without distinction'.

Court, through trading military assistance for immunity,[113] and countenancing the extradition of people to locations where they may be tortured, in clear contravention of Article 3 of the UN Torture Convention,[114] reflect an attempted realignment of what may be called 'effectiveness over morality'. Warfare may appear constrained, operating in an era of global media. However, the controls of that media are strict. Embedding of reporters and the provision of camera feeds, solely from bombs dropped from a distance on to specific targets, are examples of a continued screen over warfare. War is now fought in island camps, on airliners and in the streets of locations too remote and dangerous to have cameras. Yet it does not follow that its apparent neutrality and legitimacy are real. Rather it suggests a certain concealment has become necessary. This may be much to do with the increased curtailment of what is deemed legitimate, in part furthered by understanding the experiences of refugees and in the work of international courts. Perhaps, then, barbarization is backlash.

---

113   As permitted under the American Servicemembers' Protection Act of 2000 title II of Public Law 107-206 (22 U.S.C. 7421 et seq), see http://thomas.loc.gov, viewed on 16 June 2005.

114   United Nations, *The Convention Against Torture and Other Cruel, Inhuman or Degrading Treatment or Punishment*, 1984.

Chapter 11

'Not Being Victims Ever Again': Victimhood and Ideology

Stephen Riley

In this obsession not to become victims ever again, we allowed ourselves to become perpetrators. – Biljana Plavšić[1]

The testimony of Biljana Plavšić during her trial at the International Criminal Tribunal for the Former Yugoslavia highlighted a recurring motif in the prosecution of those authorizing and realizing abuse in conflict situations. That is, the use of real and potential victimhood as justification. Plavšić suggests that her abuses (the authorization of detention camps for non-Serbs under conditions now designated a crime against humanity) were a means of preventing the recurrence of Bosnian-Serb victimhood. This contributed to the impression that her testimony was plausible, that she deserved some mitigation to her punishment, and was a claim sympathetically received by some of the victims of her crimes.[2] It is my intention to consider why it is that these kinds of victim claims – whether they are made by 'small-fry' abusers or by powerful states – are consonant with international law and are inadequately evaluated by some contemporary philosophers.

The causal, psychological, relationship between narratives of victimhood and the authorization or realization of abuse is not something I will consider at length.[3] I would imagine that those who willingly *undertake* or actualize abuse in conflict situations have been encouraged to locate their own actions within wider, cultural, narratives of victimhood, while the relationship between victimhood and *authorization* of abuse hinges on more straightforwardly political motives. Alternatively, the motivation in both cases may be the possibility of preventing harm (to oneself/to others) through an excess of harm directed against an- or '*The*' Other.

---

1   Prosecutor v. Biljana Plavšić. ICTY Case No. IT-00-39&40/1, Sentencing Hearing, T. 610.

2   Discussion with Wendy Loweib, Victim Support Officer, Victim and Witness Unit, ICTY.

3   See Jonathan Glover, *Humanity: A Moral History of the Twentieth Century*, (Guildford, 2001).

In the case of Biljana Plavšić, I have no reason to doubt that she was, for the most part, sincere in her claims, but it seems to me that there are interesting subtexts in her admission of guilt. Admitting to having 'allowed' oneself to succumb to an 'obsession' is a subtly qualified kind of admission of responsibility: to 'allow' requires less agency that to 'choose'; an 'obsession' is possibly pathological. More specifically, there are clearly major problems with any attempt to justify pursuing a 'final solution' ('not being victims *ever* again') to the threat of violence and harm. Is it ever possible to negate all harms that you might conceivably face? Perhaps not without radical action, and perhaps even radical evil. It is interesting, and chilling, to speculate whether a successful final solution to the threat of victimhood would entail self-negation: if we managed to truly destroy all that threatened us we might well find that we depended upon that Other or Others to exist.

My present intention is, however, legal and philosophical enquiry: to try to understand the meaning of 'the victim' in contemporary international legal discourse in order to consider whether there are stable points of reference by which claims like Plavšić's can be supported or challenged. In other words, what characteristics should 'genuine' victim-claims possess? I will do this by considering those fields of international legal discourse where the role and function of the victim should be clearest – 'the humanitarian' – and place the findings into a wider theoretical context. I will conclude that international law does not provide the means for challenging victim claims like Plavšić's, that this is a consequence of the nature of international law, and that this is a problem in international law that some contemporary philosophy is ill-equipped to confront.

Victims in International Law

The language of 'victims' – victims of crimes and victims of crises – suffuses international political discourse. International law sometimes informs, but more often than not is formed by, this political discourse. It was only in the last century that the individual (as opposed to the state) began to play a role in international law, and the kinds of individuals (and conceptions of the individual) that figure in international law reflect the international political crimes and crises of the twentieth century. This is not to say that all international law (and, by extension, all international legal victim discourses) are *politicized*, but it is certainly true that international law is largely reactive and gives individuals legal personality only relatively rarely in the context of 'serious' crimes and 'serious' crises.

Victims are often invoked in the preface or preamble of legal instruments, thereby indicating that they should be thought to have an important role in those instruments. For example, victims appear early in the Geneva Conventions' Additional Protocol II:

Emphasizing the need to ensure a better protection for the victims of […] armed conflicts […].

And we find in the Preamble to the International Criminal Court's (ICC) Statute that the Parties to the Statute are:

> *Mindful* that during this century millions of children, women and men have been victims of unimaginable atrocities that deeply shock the conscience of humanity [...].

Because they appear in these preambles, victims can be held to be key animating, and therefore interpretative, principles for the instruments as a whole. This is true of the ICC Statute where 'the interests of justice' is an important and untested principle understood, in part, as the interests of victims within and outside the Court.[4] But it is worth noting from the outset the victim does not play a very substantive role *within* these instruments. For instance, there is a raft of victim welfare provisions in the ICC statute but actually very little on what constitutes a 'victim' for the purposes of prosecution and/or participation.[5] Emphasis is on the acts and intentions of the victimizer.

In more specific terms, the Declaration of Basic Principles of Justice for Victims of Crime and Abuse of Power, a General Assembly Resolution, offers the following definition of a Victim of Crime:

> 'Victims' means persons who, individually or collectively, have suffered harm, including physical or mental injury, emotional suffering, economic loss or substantial impairment of their fundamental rights, through acts or omissions that are in violation of criminal laws operative within Member States, including those laws proscribing criminal abuse of power.[6]

Two points arise from this that I wish to high-light. First, victims are equally individuals *or* collectives: this is intelligible (we often talk about victim groups), and in keeping with the scope and function of international law (dealing principally with groups and collectives). Second, as we would imagine with a legal instrument, the gauge or index of victimhood is not phenomenological (i.e. the nature, or form, of harm) but relational (i.e. arising *in relation* to acts criminalized under positive law).[7]

The victim can also be used in international law to justify the use of force, and in some instances (as Plavšić herself argued) as justification of *pre-emptive* use of force, for the prevention of the creation of victims. Recent debate over the meaning

---

4   Rome Statute of the International Criminal Court, Article 53. This provision is important because it gives The Prosecutor and Court a great deal of latitude in deciding who should be prosecuted, allowing the Court, in effect, to decide which prosecutions are most *just* or *useful* in a post-conflict environment.

5   See the 'ICC Rules of Evidence and Procedure', Rule 85 'Definition of Victims'. This sparse definition has two components: a 'victim' is a victim of a crime within the Court's Statute, and victim can include organizations and institutions.

6   Article A1.

7   This is also true of 'Victims of Abuse of Power', the second theme of this Resolution.

and significance of Article 51 of the UN Charter and the legality of the invasion of Iraq hinged on the meaning of self-defence.[8] And it would seem that there is now state practice and *opinion juris* (state acknowledgement) demonstrating that self-defence can be construed both narrowly as retaliation for acts of aggression, and in a more general way that elides regional and international stability with the broader humanitarian goals of protecting an abused victim populace.

Together, these points of reference indicate that victims can be individuals and collectives are victims of codified crimes, and that victims yield a number of different and even competing responsibilities, not least the use *and* the limitation of the use of armed force. In fact, international law is a poor resource for finding stable definitions and stable points of reference. There are at least three reasons for this. First, like any legal field it is constantly evolving and changing, and, as in any legal field, the notion of victim is relational: a victim is always a victim in relation to something else, a crime or a criminal. Second, there is much 'soft law' in the area which is useful for indicating the trajectory of law's development but which has a questionable legal pedigree and force. And, finally, because, as Martti Koskenniemi has powerfully argued, international legal arguments are ultimately reducible to two principles that must be balanced but can never be fully reconciled: state consent on the one hand, and, on the other, normativity. Or, in his terms, *apology* (for state action) and *utopia* (the drive towards binding universal norms).[9] It is arguably this equivocation at the heart of international law that gives the victim the antinomical power to both justify violence (humanitarian intervention) and to limit violence (humanitarian law). So, it may be easy to find international instruments that consider or allude to the victim, but there is actually very little concrete law surrounding victims; or, to put it another way, there is much in the way of *aspiration* surrounding victims but little of this clearly garners state *consent*.

Victims in International Human Rights Law and International Humanitarian Law

International human rights law is an area where this is clear. Human rights law broadly indicates that the violation of international human rights law creates 'victims', but the violation of these laws is not, in most instances, a crime (it does not create victims in the sense of victims relevant to a criminal prosecution). International human rights law violations *are* sometimes prosecuted where they fall under international humanitarian law (the laws and customs of law), but these remain more or less distinct spheres of legal activity with the victims of humanitarian law taking precedence in international law over the victims of violations of international

---

8   In actual fact these debates pre-date the invasion of Iraq and were mooted from 9/11 onwards. See, Steele, J., 'The Bush Doctrine Makes Nonsense of the UN Charter', *The Guardian* (London), 7 June 2002.

9   See Martti Koskenniemi, *From Apology to Utopia: The Structure of International Legal Argument*, (Helsinki, 1989).

human rights law (which would for the most part be considered to be the concern of the state). It is this divergence of human rights and the humanitarian that I wish to explore before turning to wider theoretical questions.

It has become something of an orthodoxy in commentary on humanitarian law that any further convergence of humanitarian law and international human rights law – further 'humanization' of humanitarian law[10] – would result in a backlash. Military advisors, confronted by the demand that they provide human rights in the theatre of war, would become resistant to the project of humanitarian law as a whole. This is perhaps evidenced in contemporary instances of barbarity in warfare: trained soldiers, thrust into the reality of Modern war – surgical strikes/overwhelming force used in brief risk-calculated spells/the distribution of humanitarian aid – have a blood-lust that is repressed and that occasionally manifests itself in illegal forms. So, the argument runs, when humanitarian law works, it works because it is tied to military pragmatism and the realities of actual armed conflict. Humanitarian law fails precisely where armed conflict is tempered by 'humanitarian' responsibilities: large numbers of trained soldiers being called upon not to kill but to acknowledge or even aid the enemy civilian populace.

So, while international humanitarian law and international human rights law both broadly 'protect the innocent', the pragmatism that is at the heart of humanitarian law is not really equivalent to the normative discourse of human rights. Humanitarian law is premised on the achievable (or minimal) demands of holding armed forces back from the brink of un-winnable or total war; human rights, in contrast, are a more nuanced legal discourse of ontology and normativity.

Victims and Human Rights in International Criminal Prosecutions

This contrast of international human rights law and international humanitarian law is played out in international criminal tribunals that prosecute the violation of international humanitarian law but also have some claim to be 'human rights defending' processes. Three contemporary theorists (Allott, May and Hirsh) offer different accounts of the relationship of victims, human rights and international prosecutions, and interestingly each theorist gives a different account of the role, status and function of the victim.

Philip Allott's utopian approach to international law, although it centralizes the numerous injustices that international law generates, has little to say about concrete rights and duties concerning victims. In a characteristic passage he writes, '[International law] speaks of *human rights* and means the complacent rationalizing of the systems that generate a stream of human wrongs.'[11] Critique of human rights discourse as an ideological sop that serves to prevent real justice is not new; in many cases genuine compassion for victims becomes obscured under the weight of

---

10 See Theodor Meron, 'The Humanisation of Humanitarian Law', *American Journal of International Law*, 94, 2000, p. 239f.

11 Philip Allott, *Eunomia: New Order for a New World*, (Oxford, 1990), p. 249.

their rhetoric. And, undoubtedly Allott would allow that the greater range of 'victim participation' possibilities now provided by international trials could represent a potential contribution to the 're-conceiving' of the international arena;[12] but he would also be suspicious of this 'assertion of the value of the victim' taking place in the absence of reform both domestically and internationally.

Larry May is to some extent in agreement with Allott that international criminal trials should not be seen first and foremost as a 'victory' for either human rights or the victims of international crimes:

> [T]he list of international crimes does not extend so far as to include all supposed human rights abuses, especially those that are not group-based. International tribunals should not prosecute individualized human-rights abuses as crimes against humanity. [...] International justice does not demand that victims get the convictions they request [...].[13]

May's position, which, when taken out of context, might well look rather blunt, has two merits. One, it clearly acknowledges the fault-line between human rights law and international humanitarian law: he rightly suggests that at present it is only the most egregious and systematic crimes that have clear, unequivocal, international legal status as crimes. Any attempt to treat 'human rights violations' as crimes runs up against the reality of state practice and *opinio juris* in international law. Second, it stresses that the outcome of international trials should never be a foregone conclusion. Victims are not owed justice; if anything the duty of justice lies in providing a fair trial to the defendant. In this sense May's position is a useful counterpoint to discourses of the victim and victims that overlook the challenges that achieving justice for those victims entail.

Conversely, May is perhaps too fervent in his emphasis on opinio juris and the bases of prosecution. His *leitmotif* – 'security' as the basis of normativity in the international arena – leads to 'minimalism' and 'realism'. But 'security' can be interpreted in at least two ways. First (as the common denominator between international law and international relations) security is the value of a well ordered international sphere where sovereignty is respected and where recourse to armed force features as little as possible. But that limited interpretation is often the sum total of security discourses, and was certainly the interpretation that international relations has inherited from the strategic stalemate of the Cold War. Another interpretation is offered by cosmopolitan theory: that the security of *the individual* is crucial and should not be confined to the preservation of sovereign independence.

The cosmopolitan position, along with the assertion of a universal rule of law, requires hospitality or a right of asylum: i.e. that the protection of the individual should not be confined to the provisions of the individual's native state.[14] A victim being given standing in an international court not only represents the fulfilment

---

12 Ibid.

13 Larry May, *Crimes Against Humanity: A Normative Account*, (New York, 2005), p. 254.

14 See Jacques Derrida, *On Cosmopolitanism and Forgiveness*, (New York, 2001).

of the strategic security needs of a state or states but also transcends the state in communicating the value of individuals and in contributing to a larger narrative of historical events that does not have as its starting point the state. As such, the victim has both specificity (they are given standing to present testimony that is not bounded by their relationship with a state) and is part of a wider contribution to reassertion of the value of individuals where this has been denied by systematic criminality. This, as David Hirsh plausibly argues, is not a utopian narrative of the final triumph of humanity over oppression, but a progressive move towards fully realizing the rule of law in the international arena: 'in the ICTY and the ICTR, as well as in national courts, it is coming into being.'[15]

The cosmopolitan position is correct in that the broader normative human rights discourse can and should be interpreted widely to include welfare and participation for victims. Arguing against Costas Douzinas (but just as applicably against Allott), Hirsh argues against a radical critique of 'human rights ideology' which sees the potential of human rights stultified by the limitations of international institutions:

[Douzinas] is unwilling to embrace a project of anchoring human rights, which do not have a foundation in the exclusion of non-citizens [Douzinas' critique of international law], in supra-national institutions which have some power to enforce them.[16]

Similarly, Allott is coy about the concrete provisions that can be afforded to victims of human rights abuses. But we should also, following May, sound a note of caution here: human rights and the crimes within international criminal fora remain distinct. Hirsh argues:

While humanitarian law and human rights law do have different histories, principles and purposes, it is clear that they often share the same objectives and goals; they also share a common theoretical grounding in the discourse of human rights.[17]

This position is defensible but requires two qualifications. First, although both humanitarian law and human rights law broadly 'protect the innocent', the pragmatism that is at the heart of humanitarian law is not equivalent to the normative discourse of human rights. Humanitarian law is premised on the minimal demands of holding armed forces back from the brink of unwinnable or total war; human rights are, again, a shifting discourse of law, ontology and normativity. Second, international prosecutions centralize the victim and offer them the route to publicly offer testimony (albeit mediated testimony) that contributes to the restoration of their status and to the communication of human rights values more broadly. But the reality of international courts, on which May and Allott would agree, is that, rather than contributing, through the participation of victims, to a convergence of humanitarian law and human rights law, they continue to entrench the difference between the two

---

15 David Hirsh, *Law Against Genocide: Cosmopolitan Trials*, (London, 2003), p. 159.
16 Ibid., pp. 151-2.
17 Ibid., p. 5.

(courts censure violations of the *laws of war*). And, more fundamentally, international trials function in relative isolation from the social conditions of the states from which victims come, forming a point at which *harm* is translated from one site (the state) to another (the international sphere), perpetuating some forms of harm (ethnically or racially charged conflict, amongst others), and creating other forms of harm (the subjection of testimony to doubt or alienation from dominant political discourses of restoration or forgiveness). In sum, while we can agree with Hirsh, *pace* Allot, that it would be a mistake to denounce the contribution of victims to a wider evolution of human rights in the international sphere, the failures of international law to ensure prevention of harm calls into question whether international criminal justice should be pursued where it represents the *mutation* of harm rather than the *negation* of harm.[18]

Humanitarian Intervention

If international humanitarian *law* is less concerned with victims and more concerned with achievable boundaries on the conduct of trained soldiers, the victim as a motive force in humanitarian *intervention* and humanitarian *aid* is, then, seemingly more straightforward. The 'pure' civilian – young children, women, or the elderly – can only be unequivocal victims. And this is doubly true for the victims of natural disasters. Without agency or responsibility behind their plight they seem to have unimpeachable *victim* status. In actual fact they have no clear *legal* status: humanitarian aid, and the criteria for humanitarian action, are an area of legal uncertainty. Responses to humanitarian crises are on an *ad hoc* basis and so too is the formation of the 'international community' around them. For instance, within hours of the Asian tsunami in December 2004 there was an unseemly tussle between the UN and US over who would lead the international response: either would have a legitimate claim to be representing the international community.

In sum, the victim in international law creates contradictions. The victim should be a point where legal duties are clearest, but the law is not clear; victims are given a prominent role in the interpretation of law but there is little codification of concrete duties owed to them. The victim seemingly transcends particularity (their state/their identity) with their powerful *prima facie* needs whilst, at the same time, having no common denominator or stable meaning. In other words, while the victim plays a very central normative role in international legal discourse, the actual nature of the victim remains, of necessity, unclear: it is context-specific and flexible enough to encompass both the threat and the reality of victimhood. In fact, I would argue that it is essential to the normative economy of the international arena that the victim remains ambiguous: it needs to retain its ambiguity in order to provide a flexible point of contestation and agreement. The victim is a rare point of ideological agreement in a field of profound disagreements; appeal to the victim is morally unambiguous,

---

18 See Stephen Riley, 'Harm and Transgression in International Criminal Justice', in Sönser-Breen (ed.), *Minding Evil: Explorations of Human Iniquity*, (The Hague, 2005).

but the consequences of this appeal to the victim can be wholly ambiguous. The evidence for this is abundant and overwhelming. Many pieces of international law aim to protect the victim or prevent the creation of victims. None of them – even the Genocide Convention – successfully do this.[19] I will locate this analysis in a broader theoretical context.

Victimhood and Ideology

There are very different 'paradigms' of victimhood in law and in the sciences. Science is prepared to chart the experiences of victims in countless ways; in social scientific terms, all sorts of individuals and collectives can be victims in all sorts of ways.[20] One consequence of this is that science is, unlike law, able to acknowledge that criminals themselves may be victims of some sort. For instance, psychologists might be able to identify in Plavšić's life-story intelligible (even universal) motives in her acts and in her claims; historians may be able to offer a narrative of Bosnian-Serb victimhood that makes Plavšić's acts more intelligible. Law cannot afford such niceties, rather law can be said to begrudge the status of victim. Both international law and domestic law have some guidelines concerning the identification and duties owed to victims but in the main leave victim status to be confirmed by a trial process and, even then, a trial does not really confirm victim status but only confirms the guilt of the transgressor. The victim faces a Kafkaesque problem: their experiences of victimization might be well-documented but they are never *proven*.[21] As I will argue, this is not to say that we cannot and should not use the language of the victim. And, moreover, in many instances the law is right to be cautious in affording the status of victim (an accused should be presumed innocent). But it is true that while 'the victim' finds its most rigorous and exacting criteria in the law, law generates a contradiction where this status is never truly confirmed.

This ambivalence towards the victim in law could be interpreted in terms of the work of the anthropologist Rene Girard. Girard considers the traditional role of the sacrificial scapegoat: a weak member of a community is arbitrarily chosen and killed to put an end to widespread violence, shared guilt over this victim's death generates a solidarity out of which social stability can be fostered. Consequently, Girard locates complicity in the death of innocent victims at the very foundation of law and order (providing an interesting counterpoint to more idealized social contract narratives).[22] The merit of this account is that it challenges us to see the emergence of law out of a state of nature not as a benign conferring of power but as a violent act

---

19 A rare example of 'soft' international law that is universally accepted as genuine, binding, customary international law, the Genocide Convention has failed to prevent all genocides since its unanimous adoption by the UN General Assembly in 1948.

20 In other words, individuals and collectives can be the object of a multiplicity of events and social processes that impact negatively upon them.

21 See Jean-Francois Lyotard, *The Differend: Phrases in Dispute*, (Manchester, 1989).

22 See René Girard, *Violence and the Sacred* (trans. Gregory), (London, 1988).

of killing. It also suggests that the death of innocent victims can be a powerful force for uniting people. In a globalizing international arena where inequalities of power and resources create countless victims, one of the few uniting principles in global resistance is the figure of the victim.[23]

However, I would argue that there is another, more nuanced explanation of the role the victim plays in the contemporary international arena. In the work of Slavoj Žižek we find an account of how the victim is not a shared, hidden, burden but a powerful ideological symbol, giving rise to what he calls the 'Universalization of the Notion of the Victim'.[24] Drawing on Western responses to the women and child victims of Kosovo, Žižek argues that, in the absence of political and legal certainty, 'the victim' – an ideological placeholder *par excellence* – is deployed. The figure of the victim creates an instant solidarity. But the figure of the victim creates a visceral *uninterrogated* solidarity. As Susan Sontag, writing in a similar vein, put it: 'No "we" should be taken for granted when the subject is looking at other people's pain.'[25]

Images of abuse seem to demand action, but what action: to support a cause or to commit to pacifism? What kind of solidarity do we have: with the victim or with all others who are similarly affected by the image of the victim? If our solidarity is with the 'universal victim' – the wholly innocent everywhere – how exacting is our criterion for innocence: only civilians, some soldiers, or all those on the 'morally virtuous' side of a conflict? We might put this into the context of the Abu Ghraib abuse photographs. Both the Arab World and the Western World (up to and including the US administration) expressed clear, and presumably sincere, revulsion at these images. But the consequence was business as usual: the Arab World as, *en masse*, a victim of Western brutality; the West, *en masse*, pushed into justification of victimization on the basis of the ongoing threat of terrorism. These were criminal acts, and the concrete harms suffered by the victims of these crimes have been translated into a narrative of command responsibility, the blurred boundaries of military necessity, and even the 'victims of circumstance' (as in the case of the abuser Lynndie England, 'victimized' as the 'poster girl' of American brutality). But the actual victims of these crimes, which, evidence permitting, can be afforded some kind of victim status through a trial, have been largely forgotten in a blame game concerning American victimhood in 9/11 and beyond. To put this in other terms, there has been a profound loss of particularity – the individual victims' plight – in favour of their use as interchangeable symbols of a conflict, symbols used equally to demoralize detainees or to generate moral revulsion within liberal democracies. This is victimhood-as-ideology at its purest: images of victims reduced to visceral stimulus within ideological conflict, and perhaps becoming, in turn, the catalyst of further victimization.

---

23 See Slavoj Žižek, *Iraq: The Borrowed Kettle*, (London, 2005).

24 Slavoj Žižek, *The Metastases of Enjoyment: Six Essays on Woman and Causality*, (London, 2001), p. 213.

25 Susan Sontag, *Regarding the Pain of Others*, (St. Ives, 2004), p. 6.

An ideological turn to the victim is not only found in more or less crude rationalizations of abuse, but has actually sunk deep into Western academic discourse. The work of Giorgio Agamben for instance leads the field in continental philosophical discussion of our contemporary 'post-political' situation (that is, where politics, as the art of living together, is replaced by management, the science of control), and Agamben's work finds its pessimistic apogee in the '*Muselmänner*', the walking-dead of Auschwitz. Stripped of dignity and speech, these products of the death camps are the logical conclusion of biopolitics: politics as management of the human animal.[26] The reality of the death camps becomes 'inexpressible', and the *Muselmänner*, the remnants of Auschwitz – walking, speechless, shells of existence – represent our new human condition, redeemable, if at all, only through messianic intervention.

While Agamben, Foucault and others rightly caution us to be aware of the ostensibly benign in contemporary political discourse – victim claims amongst other things – Agamben's universalization of the notion of the victim, represents a failure to distinguish ideological discourses of the victim from both the concrete realities of victimhood, and the possibilities of resistance found in law and politics. In legal, political, and baldly factual terms, *Abu Ghraib is not Auschwitz*. They are very different kinds of solutions to very different real and perceived problems and with very different victims. Scholarship should give us the means to distinguish the two: in some quarters it fails.

Does this capacity for 'the victim' to become ideological, and for scholarship to be complicit in this, undermine its normative use? In other words, if it is necessary to try and focus on the absolute context-specificity of the victim, is it therefore impossible to decide what responses to victims' legacies *en masse* should be? There are some strong arguments for emphasizing particularism here, not least because international law, for reasons outlined above, is a poor tool for generating universal normative responses to victims. It may be that 'victim' – in the international arena at least – will remain unstable and ideologically charged until other fundamental changes have taken place in the normative and political economy of the international arena. But at the same time, it is precisely because 'the victim' does have the power to generate solidarity that the concept has gained its ideological force. We cannot neglect the unifying potential of victims, but must take care that this unification is not uninterrogated, visceral solidarity, but a solidarity that is aware of both the universality and the particularity of victim claims.

Looking at the Pain of Others

In attempting to discover clear international legal principles for gauging the form and structure of genuine victim claims we hit upon two kinds of problems. First, international law cannot provide us with this because the 'victim' is one of its

---

26 See *Remnants of Auschwitz: The Witness and the Archive* (trans. Heller-Roazen), (New York, 1999).

only points of normative certainty, and 'the victim' can have this role only to the extent that the meaning of 'victim' is left unclear. Second, academic discourses of the victim are often themselves inadequate for providing more certainty. Some philosophers have treated the victim as humanity's only common denominator to the point where everyone has some claim to victim status, thereby rendering victim claims meaningless.

The shifting ground of 'the victim' and its generation of fissile formations of 'the international community' is an indication of the inability of international law to ground a stable international *rule of law*. The counter-claim that it is churlish to criticize those few remedies that international law does afford to the victim has the potential to fall foul of what Sontag and Žižek identify: the illusion that looking at the pain of others creates unity, when in fact it only *begins* to raise difficult questions concerning what we should do in the face of large-scale suffering.

The consequences of this analysis may lie in part in the fact that when warfare takes place victim claims will always be employed for any number of reasons and purposes. Some will be sincere, others not. Some will seek to limit violence, some to justify violence. International law – of necessity, by virtue of its nature, origin, and function – does not have the means to clearly distinguish between these claims. It might be assumed that academic discourse does or should have the ability to distinguish between these claims. However, it will not be able to do this effectively until it moves away from styles of theorizing that centralize and universalize the victim whilst shying away from concrete forms and patterns of victim hood. Ideology might not be eradicated from these academic discourses but they will, ultimately, be more useful than the levelling and stultifying effect of the 'universal victim'.

PART V
Making Torture Legal?

Chapter 12

Justifying Absolute Prohibitions on Torture as if Consequences Mattered

Michael Plaxton

[W]e live in a jungle of pending disasters ...[1]

In *Why Terrorism Works*,[2] Alan Dershowitz argues that American law enforcement officials should have legal recourse to non-lethal torture in situations where a figurative 'ticking time bomb' (TTB) exists, and the state needs information immediately to 'disarm' the 'bomb'.[3] This chapter does not aim to take on all (or even most) aspects of Dershowitz' argument. That is not its object because, as we will see, he refuses to make the most forceful argument he could make in favour of torture warrants – that is, the moral argument that the consequences can at times make torture morally permissible and possibly necessary. Just because Dershowitz chooses not to make this point, though, does not mean we should turn a blind eye to it. Indeed, this chapter was motivated by the view that the consequentialist moral analysis has such intuitive force that we may be asking ourselves the wrong question when we ask 'why permit torture in any circumstances?' The *real* question is 'how do we justify bright-line rules prohibiting torture outright, given the obvious stakes in at least some situations?'

The Utility of Torture

Dershowitz is not the first person to suggest that torture may have its uses. Others have claimed that torture can be an acceptable investigative strategy in rare cases

---

*   I am grateful to Carissima Mathen, Greg Gordon, David Jenkins, Fiona Leverick, and Kathryn Last for providing comments on an earlier draft of this paper. Any errors or omissions are, as always, mine alone.

1   Hunter S. Thompson, *Screwjack*, (New York, 2000), p. 22.

2   Alan Dershowitz, *Why Terrorism Works*, (London, 2002).

3   The 'ticking time bomb situation' envelops many possible hypothetical scenarios in which 'a terrorist attack ... has been set in motion so that harm to human life is imminent unless the security forces intervene so as to foil the attack.' See Amnon Reichman and Tsvi Kahana, 'Israel and the Recognition of Torture: Domestic and International Aspects', in Craig Scott (ed.), *Torture As Tort: Comparative Perspectives on the Development of Transnational Human Rights Litigation*, (Portland, 2001), p. 641n.

– the General Security Service in Israel made such an argument to its Supreme Court only a few years ago[4] – though, until recently, few in the legal academy have done so.[5] The idea that law enforcement could validly use torture as an investigative tool meets virtually universal scepticism and disapproval. That public censure has not stopped torture from taking place in a great many nations.[6] Indeed, Dershowitz takes as his departure point recent reports from American law enforcement that it may soon employ torture as a means of preventing terrorist attacks.[7]

One must resist the urge to condemn public officials for exploring torture as an option. Whatever one thinks of torture, one cannot seriously dispute its utility. Coerced confessions do not possess a high degree of reliability[8] – commentators have, at least since ancient Greece, remarked upon the manifest unreliability of admissions gleaned through torture,[9] and Dershowitz himself claims, at most, that confessions rendered through torture will not always be unreliable.[10] Ordinarily, however, one thinks of unreliability as a factor going to weight, not admissibility; that is, one does not take the unreliability of evidence as a reason to conclude that it is not evidence at all. And there may well be situations where torture would allow investigators to attain information that would otherwise remain locked away in the minds of suspects; even if torture, as an investigative tool, works imperfectly, it can provide answers not available through any other means. The 'lazy' or 'simple' investigator[11] may resort to torture for no other reason than to save herself the trouble of 'go[ing] about in the sun hunting up evidence',[12] but even the industrious and well-equipped investigator may find that no details of an imminent terrorist threat can be discovered unless she uses torture to 'unearth' it.[13]

---

4   See H.C. 5100/94 *Public Committee against Torture in Israel et al. v. The State of Israel and the General Security Service* (1999) 53(4) PD 817. For scholarly commentary on the judgement, see Reichman and Kahana, ibid.; Mordecai Kremnitzer and Re'em Segev, 'The Legality of Interrogational Torture: A Question of Proper Authorization or a Substantive Moral Issue?', *Israel Law Review*, 34, 2000, p. 509.

5   See Mirko Bagaric and Julie Clarke, 'Not Enough Official Torture In the World? The Circumstances Under Which Torture Is Morally Justifiable', (2005), *University of San Francisco Law Review*, 39, 2005, pp. 581-616.

6   See Malcolm D. Evans and Rod Morgan, *Preventing Torture*, (Oxford, 1998), chs 1-2; John Conroy, *Unspeakable Acts, Ordinary People: The Dynamics of Torture*, (California, 2000).

7   Dershowitz, *Why Terrorism Works*, p. 134.

8   See, e,g., *R. v. Oickle*, [2000] 2 S.C.R. 3 at paras. 47 ('A confession that is not voluntary will often (though not always) be unreliable.'), 69 ('The doctrines of oppression and inducements [in the confessions rule] are primarily concerned with reliability.').

9   Evans and Morgan, *Preventing Torture*, pp. 1-3, 7-8; John H. Langbein, *Torture and the Law of Proof*, (Chicago, 1976), p. 9.

10   Dershowitz, *Why Terrorism Works*, pp. 136-7.

11   Evans and Morgan, *Preventing Torture*, p. 7.

12   See Sir James Fitzjames Stephen, *A History of the Criminal Law in England*, Vol. 1, (London, 1883), p. 442.

13   See Langbein, *Torture and the Law of Proof*, p. 8, citing Piero Fiorelli, *La tortura giudiziaria nel diritto commune*, Vol. 1, (Milan, 1953-4), p. 3: 'Dreadful or not, compelling

Dershowitz' Pragmatic Analysis

This simple empirical fact – that torture can occasionally be useful and that, therefore, law enforcement agents will occasionally be tempted to use it – represents the main pillar to Dershowitz' argument in favour of torture warrants. He certainly does not appear willing to ground his argument in morality. He relies on a more pragmatic claim; namely, that the state will use torture if faced with a TTB situation, whether or not one can morally justify it. The public, he claims, would expect officials to use torture to avert imminent disaster,[14] and many of those officials would feel obligated by their office to use torture whatever their moral qualms.[15] Dershowitz assumes, on this basis, that the moral debate actually matters very little: 'The real issue … is not whether some torture would or would not be used in the ticking bomb case – it would. The question is whether it would be done openly, pursuant to a previously established legal procedure, or whether it would be done secretly, in violation of existing law.'[16] If one takes it as given that state officials will use torture then, Dershowitz argues,[17] three options are available:

> The first is to allow the security services to continue to fight terrorism in 'a twilight zone which is outside the realm of law.' The second is 'the way of the hypocrites: they declare that they abide by the rule of law, but turn a blind eye to what goes on beneath the surface.' And the third, 'the truthful road of the rule of law,' is that the 'law itself must insure [sic] a proper framework for the activity' of the security services in seeking to prevent terrorist acts.[18]

Dershowitz does not even consider the first option, which would deny that the law has anything to say regarding the use of torture by state actors.[19] He takes more seriously the 'hypocritical' view and more vigorously attacks it, arguing that one cannot reconcile that position with the need to maintain 'open accountability and visibility in a democracy.'[20] If the law proclaims that public officials may use

---

a person through violence to admit or disclose something against his will is a method of procedure so humanly obvious that it proves difficult to imagine an age in which it could not have been known.'

14 Dershowitz, *Why Terrorism Works*, p. 150.

15 Ibid., p. 151.

16 Ibid. See also Alan M. Dershowitz, 'Is It Necessary to Apply "Physical Pressure" to Terrorists – And to Lie About It?', *Israel Law Review*, 23, 1989, p. 192 at 192: 'We know, of course, what all governments would actually do [in TTB situations]: they (or more precisely, some flack-catching underling) would torture (with the implicit approval of the powers-that-be).'

17 Dershowitz borrows this analysis from the Landau Commission. See *Commission of Inquiry Into the Methods of Investigation of the General Security Service*, excerpted in *Israel Law Review*, 23, 1989, pp. 146-88 [Landau Commission Report].

18 Dershowitz, *Why Terrorism Works*, p. 150.

19 See Landau Commission Report, pp. 182-3.

20 Dershowitz, *Why Terrorism Works*, p. 152.

torture in some cases, then those responsible for moulding the law ought to be held accountable to the public. If, on the other hand, the law does not countenance the use of torture in any situation, then those who use it anyway have broken the law and the public has a right to know that. The hypocritical position deprives the public of accountability in both senses: by paying mere lip-service to the absolutist view that torture can never be appropriate (as opposed to paying genuine homage to that principle), without seriously debating the matter, lawmakers fail to take responsibility for the content of the law; and, to the extent that 'underlings' cannot publicly discuss their use of torture, they do not (indeed, cannot) take responsibility for their own investigative practices. The lost accountability of the 'higher-ups' is the more critical for Dershowitz: if someone must make a hard choice, that choice should be made 'at the highest level possible, with visibility and accountability.'[21] The 'local policeman, FBI agent, or CIA operative' does not set policy; if the law has an opinion about torture in TTB situations, then one should not expect those who merely administer investigative policies to figure out that opinion.[22]

A torture warrant regime would remove, from the hands of 'flunkies', responsibility for the use of torture, instead placing it in the hands of judges and elected officials – i.e., those who decide when such a warrant should issue, and those who actually issue them. This could conceivably result in the diminished use of torture (assuming, again, that its use is as inevitable as Dershowitz claims); a local police officer may choose not to employ torture on her own initiative if, by applying for a warrant *ex ante*, she can evade personal responsibility (in the legal, if not moral, sense) for the decision. It may, however, just as easily encourage more public officials to use torture. A professional investigator – a person who owes a duty to the public to investigate the possibility of future terrorist activities – may not feel that she can ignore the availability of a torture warrant.[23] If the investigator thinks that a court will grant her application for a torture warrant, she will experience enormous pressure to apply for such a warrant even if she would not otherwise want to use torture, since it may ultimately lead to information not otherwise attainable. If the court grants the application, the investigator will again experience enormous pressure to act on that warrant – after all, it is her job to discover terrorist threats, and information about those threats.[24] Moreover, the investigator, as only one part of an institutional regime that permitted torture to take place, can deny responsibility for administering it.[25] Investigators who would not employ torture in a system that did

---

21 Ibid., p. 154.

22 Ibid.

23 I set aside the question of whether people ought to adopt 'role moralities'. For an interesting study of the problem, see Heidi Hurd, *Moral Combat*, (Cambridge, 1999).

24 Much research has been done concerning the willingness of 'ordinary' people to torture others when their 'role' requires it. See Stanley Milgram, *Obedience to Authority*, (New York, 1974).

25 See Elaine Scarry, 'Five Errors in the Reasoning of Alan Dershowitz', in Sanford Levinson (ed.), *Torture: A Collection*, (Oxford, 2004) [volume cited as *Torture*].

not recognize its legitimate use, may frequently seek formal authorization. This, in turn, may well result in greater resort to torture.[26]

Dershowitz correctly notes that, if the front-line agent must bear responsibility for deciding whether or not to torture someone under TTB circumstances, one may not wind up with cool-headed deliberation.[27] One wonders how a torture-warrant regime would fix that problem. The TTB scenario presents grave moral problems because it anticipates that law enforcement officials will have little or no time to conduct an independent investigation and must, therefore, torture a person in order to acquire the desired information immediately. If the state has time to get a warrant, one wonders if it truly faces a TTB situation.[28] This concern gains momentum when one considers that the court hearing the warrant application will scarcely have more time to make the decision than the front-line official confronted with the problem at first instance. In making this decision, the court will need information, which means that the applicant will need time to prepare materials. If, owing to the urgent nature of the TTB crisis, the authorizing court has access only to incomplete information, such that one can anticipate erroneous decisions (whatever the terms of the torture warrant legislation), one wonders why a formal torture warrant procedure should exist. The torture-warrant regime brings little assurance that front-line agents would not wrongly torture individuals in TTB (or perceived TTB) situations; at the same time, it diffuses responsibility for the act of torture, encouraging these same agents to frequently apply for judicial authorization.[29]

Finally, Dershowitz overstates the extent to which the need for public accountability requires a torture-warrant regime. If a state actor who uses torture will face criminal charges, such that she must publicly assert a defence of necessity, that agent has been surely held accountable. One suspects that few law enforcement officials would risk a criminal prosecution unless faced with an imminent and unambiguous threat of considerable proportions; one that authorities could prevent only with information gleaned through torture. In any other case, the risk of prosecution would deter front-line officials from engaging in torture. The situation described might seem hypocritical: the state – which surely benefits when an agent uses torture to defend it – prosecutes its benefactor while declaring the evils of torture. No one, though, turns a blind eye to acts of torture. On the contrary, the state solemnly and publicly acknowledges that torture has taken place; condemns it; yet leaves room for arguments that the act of torture in question, while evil, was also necessary.

Dershowitz responds to this argument – that the prospect of a criminal prosecution generates the accountability he seeks – with a simple denial that the state would

---

26 Ibid.

27 Dershowitz, *Why Terrorism Works*, p. 162.

28 See Kent Roach, *September 11: Consequences for Canada*, (London, 2003), pp. 101-2: '[Dershowitz'] advocacy of torture is both morally and practically suspect. The examples he uses, such as a ticking nuclear bomb, are classic cases of emergency or exigent circumstances in which the police generally do not have time to obtain warrants.'

29 These points have been made by Scarry, 'Five Errors in the Reasoning of Alan Dershowitz'.

actually prosecute an agent who tortured an individual to acquire information needed to avert an imminent disaster.[30] If true, a torture warrant could accomplish little. Accountability in any legal system depends upon some prospect of public confrontation. Dershowitz claims that the state would refuse to confront agents who use torture when it is barred absolutely; if torture is not barred absolutely, and agents use torture in the absence of formal judicial authorization, one fails to see why the state would have a keener interest in confronting them. Dershowitz cannot assume that the state will act hypocritically without giving up on his argument altogether.

The Moral Argument

We cannot, then, simply set aside the fundamental moral concerns that lie at the bottom of our persistent discomfort with torture: assuming that torture is (and always is) morally unacceptable, it is hardly clear we discourage its use by devising a torture warrant regime. If torture is always wrong, we have little reason to think that the law should not treat it that way. The question, then, is whether it is always morally wrong. Unfortunately, whether we like it or not, there are good reasons to think that morality does *not* forbid torture in any and all circumstances; that absolute (or 'bright-line') prohibitions on torture are over-inclusive.

Obviously, the claim that torture is justifiable (if almost never justified) rests on a rejection of absolutist (or agent-relative) conceptions of morality. But we have good reasons for rejecting a prohibition on torture grounded in that meta-ethical tradition. An absolutist prohibition on torture will rest on the premise that the torturer's virtue is tainted by the infliction of pain, suffering, and (possibly) injury or death, but not restored by the benefits produced by engaging in the act of torture. To return to the TTB situation, saving one thousand lives cannot offset the harm inflicted upon one tortured person because the negative consequence reflects badly on the actor whereas the positive consequences make no difference. That view strikes many – perhaps most – of us as curious. It seems strange because it appears to us like the absolutist treats one class of consequences (i.e., the harm delivered upon the torture victim) as morally relevant, but not the other. If the bad consequences of torture matter in the moral sense, we intuit the good should count as well.

But, of course, to the absolutist that objection misses the mark entirely, since the thoroughgoing absolutist will claim that *neither* consequence has moral significance. The fact that the torture victim suffers pain or injury does not make the torturer's conduct morally wrong on the absolutist view; her suffering is just one factor that helps us identify the conduct *as* torture, and as therefore subject to the absolutist prohibition on it. The harm attached to the conduct operates as a moral signpost in the same way that a label attached to a thing helps us identify the thing – just as the label, though it tells us what the thing is, does not make it what it is, so the

---

30 Dershowitz, *Why Terrorism Works*, p. 162.

harmfulness of conduct, though it alerts us that a moral prohibition exists, does not make the conduct morally suspect.

To this answer, we might well ask: Why does the possibility of 1,000 people dying or suffering gruesome injury, not lead to the application of a new label (or, at least, the stripping away of the old one)? The absolutist again has an answer: because the consequences must have a certain causal relationship to the actor's conduct before they can tell us anything about its moral character.[31] Thus, the absolutist will typically distinguish relevant consequences from morally incidental consequences – for instance, by drawing a distinction between consequences directly caused from mere 'side-effects'; or between acts and omissions.[32] To the absolutist, only the agent's moral virtue matters;[33] if the consequences that flow from her conduct lack a certain causal relationship to it, they tell us nothing about her virtue and, therefore, cannot serve a signposting function.

At this point, the absolutist can only preach to the converted, for if we are not committed absolutists at the outset, we will find this account of morality radically counter-intuitive. If we think torture is morally wrong – that it is the kind of activity that law must prohibit in no uncertain terms – we surely do so because we think the consequences of torture truly horrible. We might have different consequences at the forefront of our minds: the physical harm or psychological trauma inflicted upon the victim (or, indeed, suffered by the torturer herself); or the blow delivered to the victim's dignity and self-respect. The absolutist, though, regards all those consequences as irrelevant – she denies that the *torture victim's* suffering makes any moral difference. If, the absolutist argues, we consider her suffering as a reason in itself for not torturing her, we make a serious error by acting for consequential reasons instead of moral reasons. That idea understandably offends quite a few people. (We find it all the more shocking when we realize that the absolutist is also unable to say that the lives of the thousand people who will die as a result of the TTB are morally relevant.)[34] We think human suffering is important, and we typically think morality is important insofar as it reduces the prospect of human suffering. In other words, we typically intuit that moral rules exist to advance human well-being. If they do not serve that purpose, we might be inclined to wonder if we should bother with morality in the first place.[35]

---

31 See Thomas Nagel, 'War and Massacre', in Samuel Scheffler (ed.), *Consequentialism and Its Critics*, (Oxford, 1988), p. 52 [edited volume cited as *CAIC*].

32 See Michael S. Moore, 'Torture and the Balance of Evils', *Israel Law Review*, 23, 1989, p. 280 at 298-310 (describing the various theories of culpability).

33 Ibid. at 297.

34 Some avowed non-consequentialists ('threshold deontologists') appear prepared to say that an absolutist moral rule can 'give way' in the face of catastrophic consequences. Because I believe that threshold deontology is untenable, I leave it aside.

35 See T.M. Scanlon, 'Contractualism and Utilitarianism', in Amartya Sen and Bernard Williams (eds), *Utilitarianism and Beyond*, (Cambridge, 1982), pp. 108-9 [*UB*]: 'Claims about individual well-being are one class of valid starting points for moral argument. But many people find it much harder to see how there could be any other, independent starting

Rule-Utilitarianism

If we want to justify bright-line rules prohibiting torture, we need to take consequences seriously. And, to do that, we may need to accept (however grudgingly) that torture can be useful. Perhaps it can save lives and reduce human suffering – at least on an aggregate level. In light of that fact, can a consequentialist absolutely rule out the possibility that torture could, under some horrible set of circumstances, be morally justified? Let us consider the problem, first, by considering consequentialism's most famous (or infamous) branch – utilitarianism – before discussing whether other versions of consequentialism provide greater assistance.

Any utilitarian theory, by definition, aims to maximize a 'good'. Different utilitarian theories will disagree about what should count as the good that ought to be maximized – one might say, like Bentham, that the good is pleasure; or, like Mill, that the good is human progress; whereas another might say that the good is wealth. Whatever the good is, though, it is good because it ostensibly advances human well-being either directly or indirectly.[36] The standard utilitarian line of thinking is that, if a thing (e.g., pleasure or wealth) is good, the right action must be that which will produce the most of that thing. Otherwise (as we have seen), we wind up with a morality disconnected from human interests and concerns.

Under the umbrella of utilitarianism, we find two sub-categories: act-utilitarianism and rule-utilitarianism. We will spend little time on the former. The act-utilitarian evaluates the rightness of conduct on a case-by-case basis, repeatedly and constantly nagging the actor with the question: 'will *this* act maximize utility?' If a discrete act does not, the actor has – on that occasion at least – failed to act as morality required. We can set this version of utilitarianism aside for now, since we seek here to evaluate the rightness of rules, and the act-utilitarian does not recognize the force of any rule apart from the moral rule that utility must at all times be maximized. Any rule, by definition, will be both over-inclusive and under-inclusive; that is, it will apply to situations in which its justifying rationale dictates it should not, and it will not apply to situations in which its justifying rationale dictates it should.[37] To the act-utilitarian, it cannot be right to follow a rule on an occasion when doing so fails to maximize utility, even if following the rule on most other occasions would maximize the good. If, however, we ignore the rule each and every time it leads to a result we think inconsistent with morality, then rules no longer have any power over us. That strikes us as counter-intuitive. Now, the act-utilitarian (as we will see) may be correct to think that her utilitarianism requires her to disavow the independent force of rules. But we should, at least at the outset, give utilitarianism the benefit of

---

points. Substantive moral requirements independent of individual well-being strike people as intuitionist in an objectionable sense.' See also Scanlon, 'Rights, Goals, and Fairness' in *CAIC*, p. 74.

36 Scanlon, 'Rights, Goals, and Fairness', p. 108.

37 See Frederick Schauer, *Playing By the Rules: A Philosophical Examination of Rule-Based Decision Making in Law and in Life*, (Oxford, 1991). See also Andrew Halpin, *Definition In the Criminal Law*, (Oxford, 2004), pp. 7-9.

the doubt, and suppose it can be made compatible with the authority of rules. That means we need to throw act-utilitarianism overboard.

The rule-utilitarian obviously recognizes that the quality of rule-ness can possess its own moral force. The rule-utilitarian claims we should follow rules because doing so will maximize the good in the long run. That might be because the rule, if followed consistently, charts a path towards the maximum amount of good we can possibly have at some indeterminate point in the future.[38] Were a rule-utilitarian to argue in favour of an outright prohibition on torture, she might, then, claim that the good is maximized by refraining from torture in all cases, even if it would lead to disutility in some discrete circumstances. Our trouble with that sort of rule-utilitarian justification of absolute prohibitions on torture is, of course, that exceptional circumstances (namely, TTB situations) can arise. In these circumstances, torture may result in such an extraordinary net gain in utility that it cancels out the accumulated benefits that flow from not resorting to torture in all the ordinary cases. A rule-utilitarian could plainly endorse a prohibition on torture if it incorporated an exception to account for TTB scenarios – she could choose a rule that made allowances for Dershowitz' torture warrants. There seems no reason, though, to accept on faith the rule-utilitarian's claim that an absolute prohibition on torture could be justifiable on utilitarian grounds.[39]

Alternatively, the rule-utilitarian might ground an outright prohibition in the view that it limits the *dis*utility that would result if each moral actor – hobbled by a lack of perspective, information, and impartiality – were to attempt to decide, on a case-by-case basis, what course of conduct would maximize the good.[40] Thus, we are (it might be said) better off – again, in the long run, if not in a particular case – when we ensure that police officers or military interrogators cannot weigh the merits of torture in discrete circumstances, since there is every possibility they will exercise their judgement in a non-culpable but nonetheless morally abhorrent way.[41] But this rationale makes the same assumption as its predecessor; i.e., that a more nuanced rule will lead to less disutility. We might think that police officers and military interrogators will be wrong (according to principles of utility) to use torture nine times out of ten (or 99 times out of a 100, or 999 times out of a 1000). That tenth (or hundredth or thousandth) decision to torture, though, may lead to such a staggering net increase in utility that a utilitarian justification for a less nuanced rule falls apart. The rule-utilitarian cannot, when she formulates her rules, presume that some actions are presumptively bad – after all, what *makes* conduct bad is nothing more or less than its tendency to produce disutility. She needs, then, to be able to justify her assumption that the extraordinary case will either not happen or will not

---

38 See David Lyons, *Forms and Limits of Utilitarianism*, (Oxford, 1965).

39 See Ronald Dworkin, *A Matter of Principle*, (Cambridge, Massachusetts, 1985), pp. 82-4.

40 See L.W. Sumner, *The Moral Foundation of Rights*, (Oxford, 1987), pp. 190-1.

41 I draw this distinction to recognize that a person may, with a given course of conduct, honestly and reasonably believe she is maximizing the good when in fact she is not.

offset the benefits attached to following the rule in all ordinary cases. That is a tall order, and we have little reason to think she can fill it.

So the rule-utilitarian seems unable to adopt a bright-line rule against torture. But the difficulty goes further, inasmuch as the rule-utilitarian seems unable to even say that certain *kinds* of torture, or certain kinds of victims, should be ruled out. Consider Dershowitz' own account of the benefits to be had from a torture warrant regime:

> The simple cost-benefit analysis for employing such non-lethal torture seems overwhelming: it is surely better to inflict non-lethal pain on one guilty terrorist who is illegally withholding information needed to prevent an act of terrorism than to permit a large number of innocent victims to die. Pain is a lesser and more remediable harm than death; and the lives of a thousand innocent people should be valued more highly than the bodily integrity of one guilty person.[42]

This reasoning, which certainly *looks* utilitarian, justifies the exception to the rule prohibiting torture that Dershowitz advocates. But even he is uncomfortable yoking himself to 'morality by numbers',[43] insofar as it commits him to far more than he wants to argue. Suppose we were prepared, taking consequences seriously, to permit the use of torture in extraordinary circumstances. We would *still* want to say that state agents, in TTB situations, should use the least harmful or painful methods capable of acquiring the information they seek. We would want to say there should be limits on the kind of methods available. And so, Dershowitz suggests, we can morally justify torture only if we constrain its use with rules governing when, and against whom, the state can employ it.

Unfortunately, the thoroughgoing utilitarian cannot make sense of those limits. If we are concerned with nothing more than the maximization of innocent life, or alternatively the minimization of pain, it is not difficult to imagine circumstances where this reasoning commits us to the application of *lethal* torture to one or more *innocent* persons in order to avert some disaster that would otherwise befall a larger collection of other innocent persons. If the need for information can justify interference with an individual's bodily integrity just because the state needs such information to avert a disaster to some group of people, then only four factors appear to matter in the moral calculus: the number of people in the threatened group; the nature of the threat; the number of people whom the state would require to suffer harm in order to acquire the information necessary to dispel the threat; and the nature of the harm that one must inflict. If the utilitarian cannot say that torture is *per se* out of bounds, she likewise cannot say a particular kind of torture is 'going too far'.

Dershowitz suggests that the 'prohibition against deliberately punishing the innocent' could function as a moral constraint on the utilitarian reasoning he seems to use,[44] but one wonders how. That rule basically says, 'one may not do injury to innocent persons.' The origin of this rule, in Dershowitz' theory, has an air of

---

42 Dershowitz, *Why Terrorism Works*, p. 144.
43 Ibid., 146.
44 Ibid.

mystery about it. He may consider it morally wrong to injure innocent people, whatever the consequences of doing so. But then it appears that consequences need not matter; in that case, we could just as easily resuscitate our absolutist prohibition on *torture*. If a certain meta-ethical calculus accurately determines when morality requires a particular course of action, that calculus must operate in all cases and under all circumstances or we wind up with nothing more than a moral theory of special pleading. Dershowitz must consistently apply his reasoning; if he applies consequentialist reasoning to the TTB case, he must do the same in the 'punishment of an innocent' case.

Consequentialism and Aspirations

Rule-utilitarianism, then, not only cannot give us a bright-line prohibition against torture; it is unable to say that certain methods of torture should not be used. But we can learn something from its failures, just as we can learn something from the failures of absolutism. Absolutism fails because it does not account for well-being. Utilitarianism fails because it does not account for well-being *in the right way*. In this section, we will see that, when we find a more appropriate way to account for well-being, we can also find a way to say that bright-line prohibitions on torture, if not morally required, are morally superior.

We have seen that the rule-utilitarian is vulnerable to criticism because she is forced by the terms of her moral theory to say that no rule or policy is presumptively off-limits. The moral value of an act is determined by the extent to which it adds to, or subtracts from, the total amount of good in the world. The act (or rule or policy) has no independent value. Since the act's value depends on the state of the world at the time it is performed, it fluctuates wildly – one day the value attached to torture may be low to the point of vanishing; another day, it may possess extraordinarily high value. We cannot predict what the world will be like tomorrow, never mind in a week or a year, so we cannot predict what value torture will have in the long term. This makes nonsense of any attempt, on the utilitarian's part, to justify bright-line rules prohibiting any course of action. Before we can sensibly formulate a bright-line rule barring torture, on consequentialist grounds, we need to find a way to set a value on torture that is relatively fixed. There is a way, but it requires us to measure well-being differently.

We should first note that the utilitarian's method for calculating well-being is – to say the least – suspect. We have no trouble imagining situations in which a person with more pleasure or wealth (or whatever) can be regarded as badly off, or a person with less as well-off.[45] It is, therefore, counter-intuitive to suppose that a person is better-off just because she has more 'goods'. Pleasure *might*, Nozick's thought-experiment notwithstanding, make a person well-off – consider Cipher's choice, in *The Matrix*, to be re-inserted into the matrix rather than live a 'real' life of

---

45 Consider Nozick's discussion of the experience machine in *Anarchy, State, and Utopia*, (1974).

discomfort. But Nozick had a point: many of us would prefer to live without pleasure so long as we knew we were actually doing something productive with our lives. The value of wealth, too, seems to depend on what we want to accomplish, and on whether additional wealth makes it easier for us to achieve our goals. Well-being seems to rest, not on the *amount* of 'goods' we have, but on the extent to which we, as individuals and as communities, have succeeded or failed in the worthwhile projects we undertake.[46] Thus we can, and often do, think ourselves better off when we sacrifice some 'good' (e.g., happiness or money) for the sake of the successful completion of a project we have devised for ourselves.[47]

That is an important observation because, once we peg well-being to the successful completion of worthwhile projects, we can begin to ask ourselves what projects we have, and which matter most to us. Some projects will be trivial – their success or failure will make only a marginal difference to one's well-being. Others will be so important that we will think it impossible to regard ourselves as 'well-off' if we do not succeed in them. It is worth asking if the more important projects we have include the creation of a certain kind of community – the kind of community in which, among other things, torture is not practised. If that is indeed one of the aspirations we have, not just individually but as a people, we can do what, as we just observed, the traditional rule-utilitarian is unable to do: give torture a relatively fixed and stable (dis-)value over time. We can do that because, no matter what motives the state has for engaging in torture, it acts contrary to at least one of our projects – and, so, undermines our well-being to an extent – when it does. By refusing to use torture even when the stakes are high, the state in one respect *protects* our well-being.

If our aim to become a torture-free society is important enough, it will trump all competing aims, and we can unambiguously say that an outright prohibition on torture is morally justifiable – even required – no matter the risk to our security. That, however, is a big 'if'. We may have many aspirations for our community; becoming the sort of community that does not engage in torture is probably only one. Thus, we may aspire as a community to be free of torture, yet also aspire to have a certain minimum amount of security. We would need to be quite optimistic to think that abolishing torture is so much more important to us than the other aspirations we have. More likely, we will think that the state, on some occasions, has an impossible decision to make; that success in one project means, at least temporarily, failure in another. If living in a world free of torture is not important enough to us – if our well-being does not depend *enough* on the state's refusal to engage in it 'whatever the consequences' – then the state will have options when dealing with such imminent disasters. It may reasonably decide that one project (e.g., the creation and preservation

---

46 See John Gardner, 'On the General Part of the Criminal Law', in Antony Duff (ed.), *Philosophy and the Criminal Law: Principle and Critique*, (Cambridge, 1998).

47 For example, we have no trouble recognizing that a person who wants to quit the habit of smoking is better-off when her friend withholds cigarettes, even though she has been deprived of the pleasure or happiness of smoking that cigarette.

of a secure society) is worth temporarily setting back another (e.g., the creation of a society free of torture).

So, although this consequentialist model can lead to the conclusion that an absolute prohibition on torture is morally required, it need not and probably does not. If we want to justify that kind of absolute prohibition, we will need to do a little bit more.

Supererogation and Commitment

Consequentialist morality may not require the state to abandon torture altogether. It can, however, give us the resources to criticize states that seem ready and willing to use it. Well-being is not an all-or-nothing proposition. The state has options when deciding which projects receive priority, and when deciding how many resources should be devoted to each one. So long as it acts with our well-being in mind, we cannot say it is utterly bad. We can, however, distinguish between states that do the bare minimum to support our projects, and those that do more. Consider the idea of supererogation – the notion that morality does not present a stark choice between good and bad, but between bad, good, and better. Thus, we can say that a person or institution ought to do X even if doing X means doing more than morality strictly speaking requires. The 'ought' in this usage is permissive rather than mandatory, but it still carries weight: in our day-to-day lives, if we saw a person do only the bare minimum required by morality, we might be inclined to think rather badly of her. We might, indeed, accuse her of adopting a legalistic attitude to morality – of following the letter of moral rules, rather than their spirit. That accusation might mean little to our fellows, but it can mean a great deal in the world of politics, where the object of our accusation is not a private citizen, but our elected leaders. The state truly committed to our well-being – to morality – will strive to be not just good, but better. If it settles for good when we think it could do better, we can criticize the state for adopting a miserly attitude towards morality and rights – that is, for its moral *character*. We are entitled to think that a government with low character – that is, one uncommitted to our well-being – should be replaced with a better one.

The state that uses torture in dark times, to salvage what projects it can, is not necessarily 'wicked', for we might conclude that its use of torture was morally acceptable given the implications for other important projects we have. But that does not mean it walks with the angels. We are entitled to ask how things got so bad that we had to bargain away some of our dreams for the sake of preserving others; why the state did not do more to keep us out of this situation if it is in fact committed to our well-being (for example, by doing as Kent Roach has suggested, making our water supplies and buildings more resistant to terrorist attacks and thereby decreasing the stakes of such attacks);[48] and if the state honoured our commitment to banning

---

48 See Roach, *September 11: Consequences for Canada*. Roach opines that many of the initiatives undertaken post-September 11, to the extent that they focus on preventing terrorist threats to the exclusion of policies that would reduce the impact of terrorist attacks, display

torture by using the least degrading or offensive methods of torture possible. The state that uses torture in a catastrophic emergency is not necessarily malicious – but it may be too morally disengaged to be the government we want to manage projects deeply important to us.

These judgments, in turn, can help us maintain bright-line prohibitions on torture. Michael Walzer, of course, argued there are circumstances in which politicians should be prepared to act immorally for the sake of the people they serve. We want, he argued, our politicians to have a deep respect for morality, but we do not want them so morally precious that they refuse to violate moral prohibitions in the face of catastrophic consequences.[49] One can, as a politician, have a deep respect for morality, yet be taken by surprise and compelled by force of circumstance to do things about which one feels guilt. If the guilt is genuine, though, that politician will find new strategies that permit her to keep her hands relatively 'clean' henceforth. Likewise, our state committed to well-being as its moral touchstone cannot look upon its use of torture with either moral smugness or indifference. The decision to resort to torture in exceptional cases – cases that the committed state could not have foreseen or averted – must be made *with regret*, since it is a decision to act, in at least one respect, contrary to our projects and, therefore, contrary to our well-being. Inasmuch as the state *regrets* the use of torture, it can respond to it only by changing its policies (perhaps radically, perhaps subtly) so that it need never resort to it again. The state, if it takes our well-being seriously, will craft its new policy in such a way that any future resort to torture is unforeseeable, making a warrant procedure superfluous. For this reason, torture cannot be 'normalized', e.g., through the creation of a warrant procedure – the moment the state devises such a procedure, it either acknowledges that it is not doing as much as it could to ensure our well-being, or suggests we cannot succeed in this particular project. In either case, the state must justify itself to us and takes its political chances.

There is one further sense in which we can better hold the state to account by re-conceptualizing the moral foundations of rules prohibiting torture. Debates about torture tend to be framed in all-or-nothing terms. That is a function of the two dominant ways we have of thinking about moral issues. The absolutist can only say that conduct is right or wrong – she does not acknowledge the moral significance

---

a persistent wrong-headedness on the part of politicians. One must assume, in his view, that – public officials' best efforts notwithstanding – more terrorist attacks will happen. A state will appropriately respond to the 'new' terrorist threat by accepting it as given that terrorists will successfully execute an assault on its citizens – again, despite its best efforts and not because it passively acquiesces to terrorist violence – and minimizing the possible damage or loss of life or both that can result from that successful attack. Ibid. at 21-46, 169-70, 172. Roach's suggestions are particularly interesting in light of recent plans to re-design the 'Freedom Tower' in such a way as to make it more resistant to terrorist attacks.

49 See Michael Walzer, 'Political Action: The Problem of Dirty Hands', *Philosophy and Public Affairs*, 2, 1972-3, p. 160. See also Michael Walzer, 'Political Action: The Problem of Dirty Hands' and Jean Bethke Elshtain, 'Reflection on the Problem of "Dirty Hands"' in *Torture*.

of consequences, and so cannot make sense of the claim that conduct gets 'moral brownie points' by virtue of the special consequences it brings.[50] The traditional utilitarian judges only whether a rule or policy maximizes the good. If it does, it is right. If it does not, it is wrong. Both approaches evaluate conduct in a way that permits little nuance – pass/fail. That is an unproductive way to conduct debates. It polarizes parties into those who care about consequences and those who care about moral lines in the sand. We would do better to agree that the world is worse-off when we employ torture and that we expect our leaders to devise strategies for deterring its use. The only issue is whether those leaders are as committed to stopping torture as we are. When we frame the issue as one of commitment, we can cut to the chase and ask our officials: what are *you* doing to stop torture?

Do We Really Care about Torture (and How Much)?

We can, then, hang a great deal on the idea that the moral state will advance the well-being of citizens; that well-being is advanced when people succeed in worthwhile projects; and that one of these projects is the creation of a community that does not engage in torture. Of course, we might wonder if people really do have that project. Consider the point with which we began: Dershowitz' argument that a torture warrant regime ought to exist because agents will resort to it in TTB scenarios, no matter what morality has to say about it. The prospect of many innocent civilians dying, if a guilty person is *not* tortured, will persuade many to use it. Given that inevitability, he argues, it is surely 'hypocritical' to pretend that agents would never use torture unless formally authorized to do so. Dershowitz' point might prompt us to ask two questions (neither of which are posed by Dershowitz himself). We might first ask why, if people aspire to live in a world free of torture, they seem so ready to use it. The second question, related to the first, is: even if the successful completion of worthwhile projects is central to our well-being, would people be prepared to abolish torture if it meant putting themselves at risk?

These are difficult questions, and we will not attempt a full answer here. Instead, let us say only that we *may* be able to answer both of them and that we cannot, therefore, dismiss the above argument out of hand. We might find a number of responses to the first question. Dershowitz criticizes the absolute prohibition on torture as 'hypocritical', but hypocrisy has often been described as the tribute vice pays to virtue. If there is an unhappy tendency to pretend that torture does not happen, we can attribute that to the perception that people do not want the kind of community that engages in it. (Dershowitz suggests that our objection to torture is more 'aesthetic' than moral – and, therefore, not especially compelling – but he does not say why he thinks our visceral reaction to, for instance, the dental drill scene in *The Marathon Man*, could not be grounded in our moral revulsion.)[51] True, the homage to virtue does not stop torture from happening, but this too can be explained.

---

50 See Joseph Raz, *The Morality of Freedom*, (Oxford, 1985), ch. 8.
51 Dershowitz, *Why Terrorism Works*, pp. 144, 148-9, 155.

We could point out that people may have projects, yet be unclear as to the ways in which they can succeed at them.[52] Thus, any two people with the same basic project may violently disagree about how the project can be accomplished or how best to accomplish it. We might also observe that people have more than one or two projects, and that success in one project may entail failure in another; or, less radically, that one may not be able to advance one project for a discrete period of time, without temporarily setting back another. This means that we will occasionally find ourselves forced to choose which project to advance; if we make our choices without due care, we may sacrifice our more important projects for our less important goals, and so find ourselves acting contrary to our well-being. Finally, we cannot discount the possibility that, in a culture dominated by absolutism and utilitarianism, moral actors spend less time than they should thinking about what projects they have and how important they are to them.

To the second question, as well, we might find an answer. In her critique of Dershowitz' proposal for torture warrants, Elaine Scarry raises the issue of courage. She argues that, insofar as Dershowitz assumes that law enforcement agents would not use torture to save many innocent lives unless they received formal authorization (or were confident that higher-ups would turn a blind eye), he assumes that such agents lack the courage of their convictions: they are prepared to torture a person when they can do so with impunity, but not if they face a risk of prison. Just as we might question that assumption, so we might suppose that people are willing to accept a slightly higher risk of death for the sake of building the world they want. After all, we often speak of soldiers dying for their country, of parents willing to die for their families, of settlers willing to risk their lives for a certain idea of what a country should look like. It might be telling that, in the 9/11 Commission's Report, an entire chapter was devoted to people who risked their own lives to save others.[53] We know that well-being cannot depend on us living forever. If it did, it would not really be an issue. The extent to which we are well-off depends less on our living forever than on making our lives count. Is it necessarily the case that people are prepared to assume an additional risk of death in order to create the torture-free community they want? No, but we should not assume that they would not.[54]

'The People with the *Real* Power'

People who want to take a harder line on torture than we have taken in this chapter will be uncomfortable with our reliance on the public – that is, on the public's ability

---

52 See John Rawls, 'Two Concepts of Rules', in Michael D. Bayles (ed.), *Contemporary Utilitarianism*, (New York, 1968); R.M. Hare, *Moral Thinking: Its Levels, Method and Point*, (Oxford, 1981); Sumner, *The Moral Foundation of Rights*.

53 *The 9/11 Commission Report: Final Report of the National Commission on Terrorist Attacks Upon the United States*, (New York, 2004), ch. 9.

54 On this theme of sacrifice, see Martha Minow, 'What Is the Greatest Evil?', *Harvard Law Review*, 118, 2005, pp. 2134-68.

and willingness to castigate the government for failing to demonstrate a less than full commitment to the abolition of torture. It is all well and good to talk about the projects we have as individuals and communities, but if the community is unwilling to chastise the government when it does less than it could to support them, the idea of supererogation loses its power. And, as Joseph Lelyveld recently observed in an excellent piece in the *New York Times Magazine*, the public – or, at least, the American public – has neither united to demand more information about the interrogations conducted in Afghanistan or Guantanamo Bay or Abu Ghraib, nor roundly criticized the government for the tactics employed in the interrogations they have heard about.[55] Dershowitz' predictions, that torture would neither be prosecuted nor subject to robust military censure, have proven more or less accurate. That does not mean that torture warrants will increase accountability, or reduce resort to torture, but it underscores a basic point on which Dershowitz is – like it or not – absolutely right: torture is happening, and no one is being held to account for it. Right now, the debate over 'how much torture?' and 'what kind?' is going on among elected officials, policy analysts, law professors, and moral philosophers. But the people with the *real* power – the community as a whole – seem weirdly disengaged. So what point is there in talking about their aspirations and projects, about supererogation, about the state's commitment to well-being?

It is worth observing, though, that the force of moral arguments against legal rules or executive practices has *always* depended on accountability; few have suggested that morality can exert a pull on officials merely via their respective consciences. If the public are not prepared to make officials respond to moral concerns, it is difficult to see how any moral rule or legal prohibition could make any difference to interrogation practices. We can only remind members of the public that their well-being ultimately rests in their own hands and that, if they want to put a stop to torture, they should say so. If people are not prepared to demand more information and to respond to the information they are given – well, all one can say is that we get the community we deserve.

---

55 Joseph Lelyveld, 'Interrogating Ourselves', *New York Times Magazine*, (12 June 2005).

Chapter 13

Torture, Evidence and Criminal Procedure in the Age of Terrorism: A Barbarization of the Criminal Justice System?

Dimitrios Giannoulopoulos

'The most terrible event in my lifetime … has been that torture and violence have returned … and that in some countries people are even becoming accustomed to it.'[1] These words, spoken in 1949, referred to practices occurring during a barbaric global conflict of a scale never experienced until then. Unfortunately, one could use these exact words to describe the return of torture in the beginning of the new millennium. Of course, torture had not disappeared in the decades that followed WWII, as a quick look at the annual Amnesty International reports reveals. Nevertheless, since the 1960s, torture, often-present underground, was categorically rejected by public opinion and national governments alike. The painful experiences of the recent past had rapidly led to international consensus as to the incomparable evilness of torture and opened the way towards an international law framework of *absolute* protections against it. In that respect, the mere reopening of the debate regarding the potential legalization of some form of torture could be perceived as a retreat from universal human rights values. It is certainly a paradox discussing the lawful uses of a practice that has been globally condemned as both unlawful and immoral; a practice which 'may constitute the gravest violation of human dignity and consequently of human rights.'[2]

However, torture has been at the forefront of criminal procedure legal thinking since 9/11. In the United States, Alan Dershowitz, a prominent pro-civil liberties

*   I am very grateful to Professors Andrew Choo and Geoffrey Woodroffe for looking at an early draft and for their invaluable comments.

1   F. Cocks, Speech of 8 September 1949 before the Advisory Committee of the Council of Europe, Preparatory Works : ECHR, Tome I, pp. 38-40 cited by K. Simitsis, *The Prohibition of Torture and Other Forms of Inhuman or Degrading Treatment or Punishment in International Law*, (Athens-Komotini, 1996), p. 33 (*in Greek*).

2   S. Perrakis, 'Prologue', in K. Simitsis, *The Prohibition of Torture and Other Forms of Inhuman or Degrading Treatment or Punishment in International Law*, p. 9.

Harvard Professor, has persistently[3] made the case for the use of moderate torture under exceptional circumstances and advocated the introduction of judicial torture warrants. His highly controversial thesis, based on a ticking bomb hypothetical scenario, has sparked extensive response from other academics. Most were very critical of the idea of torture warrants, while some have rejected the possibility of resorting to torture, yet being unable to deny that in some cases it might be morally justified to do so.[4] This alone is sufficient proof of the major impact that 9/11 had upon criminal procedure ideology. The imminence and intensity of the terrorist threat permitted the development of arguments which no one could have thought would be seriously debated at the dawn of the twenty-first century.[5]

Moreover, this debate was not limited to the United States. For example, two Australian academics have recently developed a Dershowitz like ticking bomb scenario, agreeing 'torture is permissible ... where the evidence suggests that this is the only means ... to save the life of an innocent person.'[6] 'Until recently it would have been hard to imagine there would be a widely-reported public debate on legalising the use of torture in Australia', reported *The Guardian*,[7] still it seems that everything is possible in the legal order after 9/11.

However, issues of torture have not arisen in the field of a *de lege ferenda* academic debate only. On the contrary, post 9/11 there is also concrete evidence of official endorsement and use of practices falling under torture. The United States have been at the focal point of criticism for resorting to torture. The images from Abu Ghraib prison are fresh in our memories. Abu Ghraib was not, however, an isolated example. The International Committee of the Red Cross found 'US interrogation techniques' in Guantanamo 'involved uses of psychological and physical coercion

---

3   Professor Dershowitz has been particularly active in advancing this argument through academic writing and the media. For analytical references to his newspaper articles and media appearances see S. Kreimer, 'Too Close to the Rack and the Screw: Constitutional Restraints on Torture in the War on Terror', *U Pa J Const L*, 6, 2003, p. 278, at p. 282, fn. 12.

4   Kreimer, for example, develops a detailed argument on why torture cannot be condoned by the Constitution, while accepting at the end that there is room for debate as to whether an official should violate the Constitution in case of a threat of mass devastation. See S. Kreimer, 'Too Close to the Rack and the Screw: Constitutional Restraints on Torture in the War on Terror', pp. 324-325.

5   The ticking bomb scenario was discussed by Bentham and other philosophers in the past. Nevertheless, it was not linked to the introduction of torture warrants. The argument for torture warrants was put forward in Israel by Professor Dershowitz himself, but rejected by the Supreme Court. Professor Derhsowitz wrote that 'before September 11, 2001, no one thought the issue of torture would ever re-emerge as a topic of serious debate in this country.' A. Dershowitz, *Why Terrorism Works*, (New Haven and London, 2002), p. 134.

6   Excerpt from paper co-authored by Mirko Bagaric and Julie Clark. Forthcoming, *San Francisco Law Review*. Cited by B. Briton, 'The Open Embrace of Torture', *The Guardian*, 1 June 2005.

7   B. Briton, 'The Open Embrace of Torture'.

that were *tantamount to torture.*[8] The *New York Times* has also reported a pattern of physical and psychological coercion followed by American interrogators after 9/11.[9] Most revealing were the memoranda of the Office of Legal Counsel in the Department of Justice, which justified the use of torture in violation of international and United States law.[10] Likewise, the method of *rendition*, otherwise known as the sending of US 'detainees to third nations where they are subjected to interrogations employing torture or other illegal techniques',[11] offers proof of official approval of torture. Vice-President Dick Cheney's recent proposal 'to allow Government agencies outside the Defense Department to mistreat and torture prisoners as long as that behaviour was part of counter-terrorism operations conducted abroad and they were not American citizens'[12] confirms, in the most emphatic way possible, US interest in rendition and torture. On this side of the Atlantic, the English Court of Appeal held that evidence obtained by torturing third parties abroad could be used in trial.[13] There are also reports according to which Britain's security agencies have relied on evidence possibly obtained by torturing detainees held in secret prisons abroad.[14]

Thus, it can be seen that post 9/11 there is a revived interest in the use of torture. In the light of the extraordinary circumstances of the 'war on terror', this interest could lead to a minimization of the absolute prohibition against torture. An important distinction needs to be drawn here. It is mainly the use of torture as a means of prevention rather than a means of establishing criminal liability that has been the subject of this modern debate on the legal use of torture. The latter has been a classic topic of the law of evidence and criminal procedure. The former was a practice unthinkable in the recent past and interest in it must be a consequence of a new

---

8   Emphasis added. J. Crook, 'Contemporary Practice of the United States Relating to International Law', *Am. J. Int'L L.*, 99, p. 264.

9   D. Van Natta Jr. et al., 'Threats and Responses: Interrogations ; Questioning Terror Suspects in a Dark and Surreal World', *The New York Times*, 9 March 2003, at A1 cited by A. Dershowitz, 'The Torture Warrant: A Response to Professor Strauss', *NYL Sch L Rev*, 48, 2003-4, p. 275, at p. 286.

10 See *Economist*, 'The Bush Administration and the Torture Memo: What on Earth Were They Thinking?', 19 June 2004, vol. 371, issue 8380; R. Goldman, 'Trivializing Torture: The Office of Legal Counsel's 2002 Opinion Letter and International Law Against Torture', *Human Rights Brief*, 12, 2004, p.1; S. Murphy, 'Executive Branch Memoranda on Status and Permissible Treatment of Detainees', *Am. J. Int'L L.*, 98, 2004, p. 820; *Time*, 'Redefining Torture', 21 June 2004, vol. 163, issue 25, p. 49.

11 J. Addicott, 'Into the Star Chamber: Does the United States Engage in the Use of Torture or Similar Illegal Practices in the War on Terror', *Kentucky Law Journal*, 92, 2004, p. 849, at p. 853.

12 Editorial, 'Legalized Torture, Reloaded', *The New York Times*, in *Le Monde, Sélection hebdomadaire de New York Times*, 5 November 2005, p. 2.

13 *A and Others v Secretary of the State for the Home Department* [2004] EWCA Civ 1123; [2005] 1 WLR 414.

14 S. Grey, 'US Accused of Torture Flights', *The Times*, 14 November 2004; D. McGrory, 'Britain Turns to *Torture Evidence*', *The Australian*, 3 August 2005.

way of global thinking in relation to counter-terrorism and national security. Both, however, are continuously shaped by recent evolutions in the field of terrorism.

The Marginalization of Evidentiary Restrictions Related to Torture

The law of criminal evidence imposes certain restrictions on the use of unlawfully obtained evidence in trial. Nonetheless, since 9/11 alternative procedures for terrorist offences have often substituted for normal criminal proceedings. This of course meant that evidentiary restrictions that are standard in criminal trials would not apply to cases of terrorism falling outside the criminal justice system. Recent developments in the United States and England allowing for more access to evidence obtained under torture illustrate this point. Before looking at such developments, I will refer briefly to evidentiary restrictions on the use of evidence obtained by torture.

Legal history is replete with examples of judicial torture to extract confessions from the defendant.[15] Nevertheless, since the eighteenth century, the common law imposed that 'involuntary confessions ... be excluded from evidence',[16] as they were considered 'untrustworthy testimony'.[17] Confessions obtained by oppression, which included torture, were deemed involuntary[18] and were subject to exclusion.

The common law rule applying to confessions obtained by oppression and torture was progressively transformed into specific exclusionary rules. In the United States, where police interrogation during the first third of the twentieth century was marked by the extensive use of the 'third degree',[19] the Supreme Court developed the *due process voluntariness test* before taking the more drastic step of the Fifth Amendment automatic exclusionary rule,[20] a constitutional rule[21] imposing the exclusion of confessions obtained as a result of a failure to administer the *Miranda* warnings.[22] The Supreme Court believed these warnings were necessary to counteract

---

15 See, for example, J. Langbein, *Torture and the Law of Proof*, (Chicago, p. 1977).

16 A. Choo, *Evidence, Text and Materials*, (1998), p. 373.

17 Expression used by Wigmore, *Evidence*, (1940), § 882, p. 246, cited by W. Ritz, 'Twenty-Five Years of State Criminal Confession Cases in the U.S. Supreme Court', *Washington and Lee Law Review*, 19, 1962, p. 35, at p. 43.

18 See *Callis v Gunn* [1964] 1 QB 495, 501; *R. v Prager* [1972] 1 WLR 260, 266 cited by A. Choo, *Evidence, Text and Materials*, p. 373, fn. 10.

19 The 'third degree' was synonymous to 'prolonged interrogation, possibly accompanied by force or threats of force.' W. White, *Miranda's Waning Protections, Police Interrogation Practices After Dickerson*, (2002), p. 14. See, also, Note, 'Judicial Torture Via the Third Degree', *North Carolina Journal of Law*, 1, 1904, p. 251.

20 See *Miranda v. Arizona*, 384 U.S. 436 (1966).

21 As recently reaffirmed in *Dickerson v. United States*, 520 U.S. 428 (2000).

22 According to *Miranda* warnings 'prior to any questioning, the person [in custody or otherwise deprived of his freedom of action] must be warned that he has a right to remain silent, that any statement he does make may be used as evidence against him, and that he has a right to the presence of an attorney, either retained or appointed.' *Miranda v Arizona*, op. cit., p. 444.

the coercion and intimidation inherent in police interrogation[23] and thus protect the right against self-incrimination. It goes without saying the admission of evidence obtained under torture violates the right against self-incrimination and is blocked by the Fifth Amendment exclusionary rule.[24]

Under English law, s.76(2) of the Police and Criminal Evidence Act 1984 (PACE 1984) prohibits the admission of evidence obtained by torture or other oppressive conduct or simply by conduct likely to render a confession unreliable. Confessional evidence can also be excluded, even when s.76(2) does not apply, on the basis of the general discretion to exclude unlawfully obtained evidence, the admission of which would have an adverse effect upon the fairness of the proceedings.[25]

Thus, in both England and the United States there are specific rules of evidence prohibiting the use in trial of evidence obtained under torture. Yet, these rules have no effect outside the criminal trial. It is particularly worrying then that in the 'war on terror' criminal proceedings are occasionally bypassed. Resorting to immigration proceedings for the purpose of detaining suspected terrorists has become standard practice. In such proceedings, the detainee is stripped of the constitutional rights she normally enjoys in a criminal trial. This may explain why immigration law has been an important tool in the fight against terrorism, facilitating the detention of suspects whom it would have been impossible to detain, or detain for a long period of time, under routine criminal justice standards.[26] Given that the evidentiary restrictions mentioned above are mainly relevant to criminal conviction rather than mere detention, these evidentiary restrictions will be considerably marginalized if the government's objective or priorities shift from criminal conviction to detention, and there are indications of such a policy shift post 9/11.

The focus on immigration proceedings and detention may result in the indirect marginalization of evidentiary restrictions. On other occasions, though, a more straightforward approach, exposing clear governmental uneasiness with standard evidentiary restrictions, is taken. For example, the Military Commission Order Number 1 of 21 March 2003, detailing the procedures for trial by Military Commissions in Guantanamo, adopted a probative value test of admissibility, allowing for the use of any evidence available to the Prosecution, regardless of the conditions under which that evidence was obtained.[27]

The war on terror can, therefore, lead to a substantial marginalization, either direct or indirect, of evidentiary restrictions. Such restrictions can be undermined in a subversive manner too. The practice of rendition is a characteristic example

---

23 For the proposition that 'the atmosphere and environment of incommunicado interrogation ... is inherently intimidating', see *Miranda v Arizona*, op. cit., p. 383.

24 See E. Griswold, *The Fifth Amendment Today*, (1955) cited by M. Strauss, 'Torture', *NYL Sch L Rev*, 48, 2003-4, p. 201, at p. 244, fn. 152.

25 S. 78 PACE 1984.

26 In that regard, see M.B. Sheridan, 'Immigration Law as Anti-Terrorism Tool', *Washington Post*, June 13, 2005, p. A01.

27 For the text of the Military Order see: http://www.defenselink.mil/news/Mar2002/d20020321ord.pdf.

of the latter. According to repeated media reports, the United States intelligence services relied extensively on this method after September 11.[28] In that way, the intelligence services may have aimed to extract confessions from terrorist suspects that they could have never obtained lawfully through police interrogation subject to evidentiary restrictions of constitutional nature. Since Bill of Rights constitutional guarantees do not have extra-territorial effect,[29] the aforementioned services could be using rendition to get round such guarantees. Thus, rendition flies in the face of constitutional prohibitions regarding the use of torture, demonstrating how far from the Constitution the government is willing to go in order to achieve its objectives.

The Court of Appeal in England legitimated a similar practice. It held that the use of evidence by torturing third parties abroad is admissible, if there is no participation of UK authorities in the obtaining of this evidence.[30] The reasoning was quite complex. First, the Court of Appeal held 'it is an abuse of process for the State to rely in any form of judicial process on the evidence of a third party or a defendant that has been obtained by torture in which the United Kingdom is implicated no matter how grave the emergency.'[31] However, it also accepted that *'the abuse of process jurisdiction does not arise if the United Kingdom authorities had nothing to do with the torture'*[32] and when the evidence was not obtained from the accused. At the same time, it held that if evidence was obtained by torturing the accused, it should be barred independently of whether the UK authorities were implicated or not.

---

28 According to the account of a US diplomat 'after September 11, these sorts of movements have been occurring all of the time.' 'It allows us,' he said, 'to get information from terrorists in a way we can't do on US soil.' Some CIA operatives say: 'Let others use interrogation methods that we don't use.' For the above, see A. Dershowitz, *Why Terrorism Works*, pp. 138, 151. For the practice of rendition, see also D. Priest and B. Gellman, 'US Decries Abuse but Defends Interrogations; "Stress and Duress" Tactics Used on Terrorism Suspects Held in Secret Oversees Facilities', *The Washington Post*, 26 December 2002, at A1 cited by J. Addicott, 'Into the Star Chamber: Does the United States Engage in the Use of Torture or Similar Illegal Practices in the War on Terror', p. 853. The New York Times has also described the existence of such a practice. See D. Van Natta Jr. et al., 'Threats and Responses : Interrogations; Questioning Terror Suspects in a Dark and Surreal World', at A1. Finally, there are even more shocking reports regarding the 'rendition' of terrorist suspects unlawfully arrested by CIA agents on the soil of European countries unwilling to co-operate with the United States, and references to secret CIA detention centres in Europe and other parts of the world. See Amnesty International, Report, *United States of America/Yemen: Secret Detentions in CIA Black Sites*, November 2005; C. Whitlock, 'Europeans Probe CIA Role in Abductions: Terror Suspects Possibly Taken to Nations that Torture', *The Washington Post*, 12 March 2005; M. Gawenda, 'An Inhuman Disgrace', *The Age*, 8 August 2005.

29 *United States v. Verdugo-Urquidez*, 494 U.S. 259 (1990).

30 *A and Others v. Secretary of the State for the Home Department* [2004] EWCA Civ 1123; [2005] 1 WLR 414.

31 See Noticeboard, 'Evidence Supplied by a Foreign State Which May Have Been Obtained Under Torture – United Kingdom (England)', *International Journal of Evidence & Proof*, 9, 2005, p. 55.

32 Ibid.

In summary, the Court's answer to the problem of admissibility of evidence obtained by torture was far from being clear and straightforward. As Deirdre Dwyer very interestingly observed:

> the court's conclusions would appear to create grades of admissibility of evidence obtained under torture, based on the degree of separation from the court of the perpetrator, the victim and the information. A confession obtained from the accused under torture by the prosecution stands at one end of the scale, and evidence obtained as a result of torture of a third party by a foreign power stands at the other.[33]

In reaching the above conclusions, the Court of Appeal turned 'a blind eye to torture by the United Kingdom's allies, and to any possible complicity by the United Kingdom in that torture.'[34] Commentators described these conclusions as 'rather shocking'[35] and 'surprising',[36] and rightly so. With all due respect, the decision seems detached from the realities associated with the 'war on terror'. The extradition of suspects to be tortured abroad is a realistic prospect. There are strong indications that it is widely used by the United States. Yet the Court of Appeal failed to put a disapproval stamp on the use of evidence obtained under torture abroad. The great risk of unfairness to the defendant who is confronted with evidence obtained by the torturing of a third party that he cannot cross-examine, the potential unreliability of such evidence as well as the effect that judicial condoning of torture can have upon the integrity of the criminal justice system should have allowed the Court of Appeal to send a strong condemnatory message. It is therefore hoped that the House of Lords will dissociate the English judiciary from judicial torture.

To sum it up, the use of alternative proceedings to try terrorist offences and the use of methods like rendition or torturing third parties abroad to obtain confessions and information against suspected terrorists are indications of disrespect towards fundamental evidentiary restrictions and traditional criminal justice guarantees. They simply demonstrate there are no constitutional or fair trial boundaries that cannot be exceeded in the 'war on terror'. They show that the ends, no matter how appalling, justify the means. In practical terms, evidence obtained by torture could now be admitted in criminal trials in both the United States and England, if obtained abroad under specific circumstances, while its use in immigration and other non-criminal proceedings also falls outside the scope of the exclusionary rules. Developments since 11 September may thus pinpoint particular hostility on the part of the executive towards evidentiary restrictions regarding the use of unlawfully obtained evidence. One could therefore say we might be moving *de facto* towards

---

33 D. Dwyer, 'Closed Evidence, Reasonable Suspicion and Torture: *A and Others v. Secretary of State for the Home Department*', *International Journal of Evidence & Proof*, 9, 2005, p. 126, at p. 131.

34 Ibid.

35 B. Dickson, 'Law Versus Terrorism: Can Law Win', *EHRLL*, 1, 2005, p. 12, at p. 27.

36 A. Mukherjee, 'Court of Appeal – Special Immigration Appeals Commission: Admissibility of Evidence Obtained by Torture', *JoCL*, 69, 2005, p. 16.

new procedural standards for terrorist offences that, amongst other things, would also make accessible in trial all available evidence, regardless of the way in which it was obtained. This movement is implicit in the marginalization of evidentiary restrictions described above. It becomes much more obvious in the debate about the legal use of torture to prevent the commission of terrorist offences.

Judicial Torture to Prevent the Commission of Terrorist Offenses

In the wake of 11 September, the much-discussed academic exercise of the 'ticking bomb' scenario re-emerged. One could assume that it was the heightened risk of a ticking bomb situation actually occurring that sparked the debate about the authorized use of torture in such circumstances. This may explain how the classic ticking bomb scenario was connected with the idea of judicial torture warrants, which will be the focus of this chapter.

In his influential book *Why Terrorism Works*, Alan Dershowitz advocates the introduction of torture warrants that would permit the torturing of a suspected terrorist in a ticking bomb case. He refers to three versions of the ticking bomb scenario. In all three of them, a suspected terrorist is held by the police. The police know he has planned a terrorist attack and the attack is imminent. The suspect will not reveal details of the attack that could prevent it from taking place and the only available means is the infliction of torture to make him talk. The question is whether the interrogators should be allowed to inflict torture under such circumstances.

In these three versions, the calculated damage is different. In one of them, bombs planted across city buildings will explode.[37] The second refers to the attack of 9/11. In the third one, a nuclear bomb is set to explode killing hundreds of thousands of people. All three attacks are imminent, but in the first version the bomb will explode within the next 24 hours, in the second the attack *appears* to be imminent, while there is no precision as to the imminence of the attack in the third version. It is equally uncertain that the suspect knows about the location of the bomb in one of the scenarios. In another, the interrogators are not sure torture will work. In spite of these differences, Dershowitz gives an unexceptional answer to the ticking bomb dilemma. He argues that under such circumstances 'the simple cost-benefit analysis for employing ... non-lethal torture seems overwhelming: it is surely better to inflict non-lethal pain on one guilty terrorist who is illegally withholding information needed to prevent an act of terrorism than to permit a large number of innocent victims to die.'[38] Having resolved the 'choice of evils' dilemma, he then goes on to make the case for the legalization of torture through the use of judicial warrants.

---

37 This is the scenario developed by philosopher M. Walzwer, 'Political Action: The Problem of Dirty Hands', *Philosophy and Public Affairs*, 1973. Alan Dershowitz starts his ticking bomb discussion with this version. A. Dershowitz, *Why Terrorism Works*, p. 140.

38 A. Dershowitz, *Why Terrorism Works*, p. 144.

Under such a system, torture would be authorized only if judges found compelling evidence obliging them to issue a warrant.[39]

Unfortunately, Dershowitz offers no further enlightenment as to the exact nature of the circumstances that would justify issuing a torture warrant. As the three ticking bomb variations mentioned above show, these circumstances could cover very wide ground. Damage, imminence, amount of information possessed by the suspect, alternatives to torture, chance of success if torture inflicted and degree of officials' certainty about all the above are amongst the factors that one should need to explore before being able to advocate the implementation of torture warrants. Otherwise, the 'slippery slope' effect cannot be avoided. Dershowitz recognizes the problem with the slippery slope argument, but simply argues that 'an appropriate response … is to build in a principled break.'[40] He does not go into the details of such a principled break though. Doing so could have been fatal for his argument. In that regard, Strauss was right to observe that 'numerous questions arise that make it impossible to even determine the exact scenario Dershowitz imagines.'[41] The factors mentioned above could account for unlimited variations of what could be labelled a ticking bomb scenario. Therefore, one could argue that the ticking bomb scenario is misleading.[42] It is used in order to make the case for torture warrants, while, in reality, it is practically impossible to define when such warrants could be issued.

To base the introduction of torture warrants on the ticking bomb scenario is thus certainly problematic. There is substantial asymmetry between the two main elements of the argument. In particular, there is no logical connection between the answer to the problem of what should happen in a ticking bomb situation and the proposed solution of the legalization of torture as a means of prevention. Just because it *might* be *justified* to use torture under truly exceptional circumstances, it does not follow that the judicial system should authorize the infliction of torture. There is much ground to be covered and many obstacles to be overcome before one could achieve a successful transition from the ticking bomb to torture warrants.[43]

Dershowitz attempts such a transition through accountability. Despite extensive reference to the ticking bomb scenario, his main preoccupation is not with the situation described under this scenario, but mainly with the fact that torture is being inflicted underground anyway. As he notes himself, the torture debate 'is not so much about the substantive issue of torture, as it is over accountability, visibility,

---

39 Ibid., p. 158.

40 Ibid., p. 147.

41 M. Strauss, 'Torture', p. 265.

42 As Clive Stafford Smith noted, the ticking bomb is 'a myth used to justify a nightmare'. C. Stafford-Smith, 'Torture is Rife Because our Leaders Encourage it', *The Independent*, 26 February 2005.

43 As Bob Brecher argues, 'the individual responses to such moral dilemmas [the ticking bomb scenario dilemmas] are one thing, the required professionalization of torture quite another.' B. Brecher, 'Torture is Always an Evil Option', *The Times Higher Education Supplement*, 18 June 2004, p. 15.

and candour in a democracy that is confronting a choice of evils.'[44] He believes that 'at least moderate forms of non-lethal torture' 'are in fact being used by the United States ... today.'[45] He wants to see this phenomenon eclipse and believes that 'a formal requirement of a judicial warrant as a prerequisite to non-lethal torture would decrease the amount of physical violence directed against suspects.'[46]

Again, one can identify the irrationality of connecting the ticking bomb situation with torture warrants. The former is an extreme situation of high visibility that 'almost certainly, will never happen.'[47] The latter are theoretically contemplated as an appropriate response to torture that is widespread and happening below public visibility. On top of that, the accountability-based argument for torture warrants would be problematic even if it were not associated with the ticking bomb scenario. No matter how noble the objective of fighting torture that occurs secretly, the method proposed to achieve this objective is unreasonable. If the problem is that torture goes unchecked, then the discussion should be about how to identify torture and punish the offenders accordingly. If the main objective is to diminish torture, it is paradoxical to suggest that torture is legalized. Instead, additional safeguards should be provided so that we can prevent torture from infecting the criminal justice system. Dershowitz' effort to enhance accountability for police conduct during interrogation should be applauded. It is the means he has chosen in his effort to do so that is questionable.

It is submitted here there are viable accountability and control alternatives to a degrading endorsement of torture. Video and audio recording of police interrogation, extensively used in England and recently introduced in France, is one of them. A more active role for legal counsel during interrogation is another. The conduct of interrogation under the strict control of a judicial officer, which is common in European continental legal systems, is alien to common law interrogation practice, but should be seriously considered if elimination of judicial torture is to be realistically attempted. A harsher stance should be equally taken in regard to criminal punishment or civil liability of officers convicted of torture. Finally, there could be guarantees external to the criminal justice system, like the institution of an independent 'Police Ombudsman' with enhanced power to control the conduct of police investigation in accordance with constitutional and other procedural guarantees, to report on any violations identified, and to take an active role in the related disciplinary proceedings. Therefore, it is possible to achieve police accountability without adopting undemocratic methods such as torture. Torture should not even be seen as a last resort to enhance accountability, let alone be presented as the sole possibility in that direction when other possibilities have not been explored.

It has been demonstrated above, first, that the 'ticking bomb' scenario is irrelevant to the discussion about 'torture warrants' and second, that the objective

---

44 A. Dershowitz, 'The Torture Warrant: A Response to Professor Strauss', p. 279.
45 Ibid., p. 277.
46 A. Dershowitz, *Why Terrorism Works*, p. 158.
47 See M. Strauss, 'Torture', p. 270.

of accountability cannot legitimate judicial torture. The more general case against judicial torture as an interrogation tool to prevent terrorist offences will now be made.

The success of the hypothetical 'ticking bomb' scenario is based on a utilitarian argument according to which the benefits of inflicting torture in a particular case far outweigh the costs of omitting to do so. Jeremy Bentham made a utilitarian argument for torturing one person when this would save the innocent lives of a hundred persons[48] and Dershowitz built on that, saying torture would certainly be 'justified to prevent the murder of thousands of innocent civilians in the ticking bomb case.'[49] This is indeed a very strong argument that cannot and most probably would not be neglected, if such a scenario were to arise in reality. Nevertheless, this is far from justifying the introduction of torture warrants.

First of all, the utilitarian argument has no limits. If a cost-benefit analysis can lead to the legalization of torture to save thousands of lives, or hundreds of lives, why not legalize it to save dozens of lives, or to save the life of one innocent person rather than protect the bodily integrity and human dignity of one allegedly guilty person? Or, why not use torture upon an innocent person to save the lives of thousands/ hundreds/dozens of innocent persons or simply the lives of two innocent persons? Answers to these questions become incredibly complex as different variations are presented. One could keep on imagining similar situations, since the legalization of torture to prevent a terrorist attack would open Pandora's box. The problem is that the 'ticking bomb' scenario was stated in a way that would make the case for the use of torture as plausibly as possible. Yet, there are innumerable 'ticking bomb' scenarios where the use of torture would be highly contested, if not categorically rejected.

More importantly, criminal procedure, even in its most basic format, is not compatible with an 'ends justify the means' ideology that is inherent in the utilitarian argument behind the ticking bomb scenario. One could refer to the German Supreme Court, which stated 'there is no principle in criminal procedure law according to which any means can be used in the pursuit of truth.'[50] In fact, criminal procedure is the exact opposite of 'ends justify the means' ideology, since it most often involves restricting the means available to the police or the prosecution service in order to protect values alien to accurate adjudication and conviction of the guilty.[51] Likewise, one should not forget it is 'ends justify the means' reasoning that underlies terrorist activity anyway: the acceptance that the infliction of death upon innocent persons is a justifiable means in the direction of achieving certain political or financial ends. Accepting the use of torture to achieve a certain objective, no matter how worthy

---

48 See W. Twining – P. Twining, 'Bentham on Torture', *Northern Ireland Legal Quarterly*, 1973, p. 347 cited by A. Dershowitz (2002), *Why Terrorism Works*, p. 143.

49 A. Dershowitz, *Why Terrorism Works*, p. 143.

50 BGHST, tome 14, p. 365.

51 For a very interesting description of such values, see A. Ashworth, *Human Rights, Serious Crime and Criminal Procedure*, (London, 2002), p. 9 ss.

that might be, would make states living under the terrorist threat no different from the terrorists,[52] it would cause these states to lose their 'moral high-ground in the battle against terrorism.'[53] Yet, maintaining moral legitimacy when fighting the 'war on terror' is of fundamental importance.

In addition to that, a Western democracy openly legitimating torture would set a very poor example for others to follow.[54] If torture warrants were introduced in the democratic United States, what could stop African or Middle-Eastern dictatorships from using torture? To take this further, one could refer to the coordinated effort in the United States to dignify the war in Iraq by stressing the benefits of the ongoing democratization process. Endorsing torture back home while 'fighting for democracy' in Iraq is hypocritical. The United States and any other country fighting the 'war on terror' have to adopt a sincere democratic policy, if they are to persuade the rest of the world about their democratic agenda.

The endorsement of torture would not only send a self-contradictory message to the world, but to the police as well. The exclusionary rule, for example, serves to deter police from resorting to unlawful conduct. Torture warrants would send the conflicting message that police can, under specific circumstances, apply torture, let alone adopt other coercive methods.

Likewise, there are important side-effects to the legalization of torture that have been generally neglected in the relevant debate. By using torture, we might be running the risk of new and more violent attacks by terrorist groups willing to avenge the infliction of torture upon their members as well as to demonstrate their determination to achieve their objectives. Furthermore, we might be making martyrs out of tortured terrorists, thus helping the existing terrorist groups generate new breeds of terrorists all over the world.[55] In addition to that, as the London attacks equally prove, we already face a new breed of suicide bombers. Using torture could only make it more likely that suicide bombings will become standard, as terrorist groups might think the potential infliction of torture could lead police authorities directly to them. Finally, from another perspective, there is the serious risk of 'police laziness'. Allowing the police to apply a *prima facie* very efficient means of interrogation like torture might dissuade them from using less efficient means of interrogation, like acceptable police trickery and promises, or from making the effort to achieve cooperation with the terrorist in exchange for potential benefits.

It has to be underlined here that the above arguments mainly go against the implementation of 'torture warrants', rather than the ticking bomb scenario itself.

---

52 See also J. Addicott, 'Into the Star Chamber: Does the United States Engage in the Use of Torture or Similar Illegal Practices in the War on Terror', p. 911.

53 O. Gross, 'Are Torture Warrants Warranted? Pragmatic Absolutism and Official Disobedience', *Minn. L. Rev.*, 88, 2004, p. 1481, at p. 1505.

54 See, in that respect, B. Boutros-Ghali, 'US Torture Sets Back Cause of Human Rights in Arab World', *New Perspectives Quarterly* 21(3), 2004, p. 9.

55 As Boutros Boutros-Ghali observes, photos from Abu-Ghraib 'are a gift to Al-Qaeda and to other terrorist groups that will be formed in the future'. See B. Boutros-Ghali, 'US Torture Sets Back Cause of Human Rights in Arab World', p. 9.

Torture warrants could lead to the standardization of torture. A state could suffer the loss of its moral legitimacy or send the wrong message to the world about its adherence to democratic principles if seen as generally endorsing torture, and not so much if seen as only replying to an emergency situation of extreme gravity threatening its existence, like in an apocalyptic nuclear bomb scenario. Likewise, there would need to be a pattern of torturing suspected terrorists before any side effects might occur.

The paradoxical nature of the torture debate compels one to focus on such policy considerations as mentioned above. As a result of post 9/11 utilitarian legal thinking, one is forced to demonstrate that legitimating the infliction of torture would be disastrous on a practical level. On a normative level, one would not need to do this though. The legal prohibitions against torture are overwhelming. From a national law perspective, constitutional charters in nearly every civilized nation in the world unexceptionally condemn torture. The use of torture is equally described as a serious criminal offence worldwide. It is in international law though one finds a complete framework of categorical prohibitions of torture.[56] In fact, unlike other international human rights, there is no deviation from the right against torture, not even in times of war or public emergency.[57] The right is absolute. Therefore, given that the legal argument against torture is so strong, the torture debate raises very important questions about our commitment to the rule of law. The mere fact that this chapter needed to concentrate on policy considerations rather than an analysis of the law proves that in the 'war on terror' we have long crossed the limits of legality. This chapter simply adds that we now seem equally prepared to overstep the boundaries of reason, alienating ourselves from the core humanistic values we have long cherished and fought for.

Judicial Torture and Legal Evolutions after 11 September

If the criminal justice policy and legal arguments against torture warrants are so compelling, it is very perplexing that distinguished authors advocate their introduction, and equally problematic that others simply consider them an 'unfortunate necessity'[58] in the 'war on terror'. To achieve a better understanding of the roots of the debate, it thus becomes indispensable to examine judicial torture against the background of criminal procedure evolutions after 9/11.

---

56 Article 5 Universal Declaration of Human Rights; Article 7 International Covenant on Civil and Political Rights; Article 3 European Convention on Human Rights; Article 4 Charter of Fundamental Rights of the European Union; Article 5(2) American Convention on Human Rights; Article 5 African Charter on Human and Peoples' Rights; UN Convention Against Torture.

57 In that respect, see, for example, article 15 ECHR. See also J. van der Yver, 'Torture as a Crime under International Law', *Albany Law Review*, 67, 2003, p. 427, at p. 453.

58 See, for example, S. Levinson, 'Precommitment and Postcommitment: The Ban on Torture in the Wake of September 11', *Texas Law Review*, 81, 2003, pp. 2048-9.

Anti-terrorism legislation introduced in the United States and England in the wake of 9/11 led to serious compromise of fundamental human rights values inherent in the rights-based theory of the criminal process.[59] Fundamental guarantees like the right against unreasonable search and seizure or against electronic surveillance in violation of the right to privacy, the right against self-incrimination and the right to counsel or the presumption of innocence, to name but a few, were severely curtailed. This was parallel to a significant undermining of judicial control within the criminal process. Moreover, post 9/11 there was heavy reliance on detention without trial and a general trend of dealing with terrorist cases outside the criminal justice system. In the limited cases where criminal proceedings were opened, one could observe a considerable number of exceptions alien to the criminal trial, like limited access to evidence and witnesses. Likewise, such criminal proceedings were often taking place under the threat of a stay, followed by a possible transference of jurisdiction to military commissions, if things did not evolve well for the prosecution.

Therefore, one debating judicial torture should take into account that after 9/11 the criminal justice system saw many innovations previously unimaginable and in stark contrast to a human rights approach towards criminal process values. For example, secret searches (*sneak and peak*), introduced with the USA PATRIOT Act,[60] constituted a serious violation of the Fourth Amendment. Likewise, extended surveillance possibilities in both England and the US were not blocked by the right to privacy.[61] Another example from US law was the Attorney-General's Regulation permitting the monitoring of attorney-client communications,[62] which was a serious breach of the fundamental constitutional right to counsel under the Sixth

---

59 For reviews of such legislation, see H. Fenwick, 'The Anti-Terrorism, Crime and Security Act 2001: A Proportionate Response to 11 September?', *Modern Law Review*, 65, 2002, p. 724; C. Harding, 'International Terrorism: The British Response', *Singapore Journal of Legal Studies*, 2002, p. 16; P. Heymann, 'Civil Liberties and Human Rights in the Aftermath of September 11', *Harvard Journal of Law & Public Policy*, 25, p. 440; D. Suchar, 'Panel Discussion – The USA PATRIOT Act and the American Response to Terror: Can We Protect Civil Liberties After September 11?', *American Criminal Law Review*, 39, p. 1501; A. Thomas, 'September 11th and Good Governance', *Northern Ireland Legal Quarterly*, 53(4), p. 366; A. Tomkins, 'Legislating Against Terror: the Anti-Terrorism, Crime and Security Act 2001', *Public Law*, 2004, p. 205.

60 Article 213 PATRIOT Act. Nevertheless, Congress later revisited this provision and proceeded to indirect abolition. See The New York Times, 'House Takes Aim at Patriot Act Secret Searches', 22 July 2003.

61 See J. Evans, 'Hijacking Civil Liberties: The USA PATRIOT Act of 2001', *Loyola University of Chicago Law Journal*, 2002, p. 933; S. Rackow, 'How the USA PATRIOT Act Will Permit Governmental Infringement Upon the Privacy of Americans in the Name of "Intelligence" Investigations', *University of Pennsylvania Law Review*, 150, p. 1651; A. Tomkins, 'Legislating Against Terror: The Anti-Terrorism, Crime and Security Act 2001', pp. 209-210.

62 Code of Federal Regulations 501.3(d).

Amendment.[63] Furthermore, detention of terrorist suspects outside US jurisdiction led to a *de facto* suspension of habeas corpus, while the UK Government was even more forthcoming than the American one in that respect, going so far as to opt out of article 5 of the European Convention in order to implement detention without trial.[64] The UK Government was forced to scrap indefinite detention after strong condemnation by the House of Lords,[65] but then proceeded to introduce equally problematic measures like house internment.[66] In general, both the US and UK Governments were 'not slow in coming forward with new legislation extending the powers of the state in many directions',[67] legislation that could have a major effect upon the character of the criminal justice system.

Therefore, one could assume that in the light of this radical degradation of values inherent in the rights-based criminal justice model the legalization of torture was seen as a realistic possibility in the war on terror, in spite of the obvious legal and moral objections against it. The emerging consequentialist approach to the war on terror and post 9/11 politics of fear made it possible to advocate the legalization of torture, a practice wholly incompatible with the rights-based approach to criminal process values. The serious undermining of so many fundamental procedural guarantees following 11 September may have acted as an avalanche, destroying in its path the credibility and efficiency of even the most sound of constitutional principles.

Be that as it may, one should not lose sight of the specific evilness of judicial torture. The torture debate may have been facilitated by a parallel surrender of universal criminal justice values after 9/11, but because of the utterly barbaric nature of torture, the torture debate possesses a unique content and its message is far more brutal than that attached to other similarly problematic procedural measures.

All in all, the debate on the legalization of torture, coupled with a substantial departure from the standard model of criminal proceedings after 9/11, reveals particular hostility and uneasiness with the rights-centred model of criminal procedure. It also denotes an underground transition towards a new procedural model for terrorist offences. Simply looking at legal evolutions within the criminal justice system after 11 September, one would not find it difficult to imagine the characteristics of such a *sui generis* model: ultra-efficient investigative methods, *including torture*; no, or simply very basic, defendants' rights and absolute freedom

---

63 See, in that regard, comments made by the President of Defense Counsels in the United States, Irwin Schwartz, in *Detroit Free Press*, November 10, 2001. See also A. Amar & V. Amar, 'The New Regulation Allowing Federal Agents to Monitor Attorney-Client Conversations: Why It Threatens Fourth Amendment Values', *Connecticut Law Review*, 34, 2002, p. 1163.

64 See J.L. Black-Branch, 'Powers of Detention of Suspected International Terrorists Under the United Kingdom Anti-Terrorism, Crime and Security Act 2001: Dismantling the Cornerstones of a Civil Society', *E.L.Rev. Human Rights Survey*, 27, 2002, p. 19; A. Tomkins, 'Legislating Against Terror: The Anti-Terrorism, Crime and Security Act 2001', pp. 210-217.

65 *A. v Secretary of State for the Home Department* [2005] 2 WLR 87.

66 See 'Control Orders' under the Prevention of Terrorism Act 2005.

67 A. Ashworth, *The Criminal Process*, p. 15.

of proof, no judicial control over police conduct, heavy reliance on detention without trial and pre-emptive detention, secrecy and absence of lay participation in the administration of criminal justice as well as an active role for the intelligence services and close international cooperation.

Such a model does not lie in the sphere of imagination. In any case, specific anti-terrorism legislation has existed for a long time in countries living under the terrorist threat, the United Kingdom, France, Italy and Germany being characteristic examples.[68] This legislation was introducing exceptions to procedural standards applying to criminal cases. Yet, since 9/11, national governments were often not limited to that. On the contrary, by implementing Draconian anti-terrorism measures and by bypassing established criminal justice procedures, they often led to a complete invalidation of general procedural standards as opposed to a minimization of their effect through the provision of certain exceptions. After 9/11, there was a major policy shift from applying specific versions of criminal procedure standards to bypassing criminal proceedings altogether and adopting methods incompatible with any notion of justice. This is why one could now identify a possible movement towards a *sui generis* model for terrorist offences.

This policy shift was presented as the result of the changing nature of the fight against international terrorism and the particular gravity of the terrorist threat. The global character of the latter compels countries under the terrorist threat to resort to international cooperation. Therefore, given terrorism's 'capability' as 'a driving force for convergence',[69] this *sui generis* model for terrorist offences could become an internationally homogenous procedural model. Francis Pakes identifies two possible ways in which terrorism can constitute a convergence factor. There is the possibility, he argues, of simultaneous development, which simply means similar responses to similar threats.[70] There is also the possibility of internationally driven convergence as a reflex to the globalization of crime,[71] in our case terrorism. Both possibilities now seem very realistic. If one looks at anti-terrorism legislation passed since 9/11 in countries facing a serious terrorist threat, one will be surprised by the number of similarities between them, which can be either the result of a common sense as to an identical threat or the product of internationally driven harmonization.

After the events of 7/7 and the recent escalation of terrorist attacks all over the world, inter-state agreements and international initiatives regarding the fight against terrorism have increased considerably. This trend is likely to persist, as 'the search for a *droit commun* ... seems inevitable in a world where the relations of interdependence linking all human beings, and even the future generations, to

---

68 See M. van Leenwen, 'Democracy Versus Terrorism: Balancing Security and Fundamental Rights', in M. van Leenwen (ed.), *Confronting Terrorism: European Experiences, Threat Perceptions and Policies*, (The Hague – London, 2003).

69 F. Pakes, *Comparative Criminal Justice*, (2004), p. 176.

70 Ibid., p. 177.

71 Ibid., pp. 177-178.

an involuntary society of risks, have multiplied.'[72] In this international society of universally shared risks, where human beings are interdependent when fighting for 'survival' in the age of terrorism, convergence of anti-terrorism legislation now seems more likely than ever before. This is very worrying. Since World War II, criminal procedure systems have been converging on the basis of mutual respect towards core universal human rights,[73] as a result of the '*constitutionalization/conventionalization* of criminal procedure' phenomenon.[74] The latter is the progressive integration of constitutional or international convention principles within criminal procedure. The constitutionalization of American criminal procedure under the Warren court and the steady incorporation of ECHR procedural principles in continental European legal systems, and more recently in the United Kingdom as well,[75] are demonstrations of this phenomenon and explain a certain amount of convergence between national criminal justice systems on both sides of the Atlantic. Yet, if convergence is now likely to proceed on the axis of a common terrorist threat, this could lead to a *reverse evolution of criminal procedure*, given the annihilating effect of the terrorist threat upon civil rights ideology.

Taking into account the interconnections between legal developments after 9/11 and the torture debate, one could say such a reverse evolution might become synonymous to a barbarization of the criminal justice system. However, a potential barbarization of the criminal justice system can strike one as even more surprising and upsetting a phenomenon. Barbarity is inherent in warfare and international law is an attempt against the odds to bring into it principles of humanity. Criminal procedure, on the contrary, is *de facto* a humane and dignified system of determining criminal liability. The smallest form of barbarity is antithetical to its very nature.

---

72 M. Delmas-Marty and M.L. Izorche, 'Marge nationale d'appréciation et international-isation du droit – Réflexions sur la validité formelle d'un droit commun en gestation', in Unité Mixte de Recherche de Droit Comparé de Paris, *Variations autour d'un droit commun – Travaux préparatoires*, (Paris, 2001), p. 74 citing J. Habermas, *La paix perpétuelle, le bicentaire d'une idée kantienne* (1997).

73 As Christoph Safferling notes, despite their differences Anglo-American and Continental legal systems 'coexist under the same human rights concept'. C. Safferling, *Towards an International Criminal Procedure*, (Oxford, 2001), p. 366.

74 For a study of the constitutionalization/conventionalization of the law of evidence and criminal procedure see K. Martin and F. de Melo e Silva, 'La constitutionnalisation/ conventionalisation du droit de la preuve', in G. Giudicelli-Delage (dir.), H. Matsopoulou (coord.), *Les transformations de l'administration de la preuve pénale. Perspectives comparées: Allemagne, Belgique, Canada, Espagne, États-Unis, France, Italie, Portugal, Royaume-Uni*, (2005).

75 In that respect, see Markezinis who refers to 'the Europeanization of English law', adding that 'French, German, Italian, and other continental systems have been experiencing similar convergence prompted by Supra-State factors such as the activities of the two European Courts.' See B. Markezinis, 'Learning from Europe and Learning in Europe', p. 30 in B. Markezinis (ed.), *The Gradual Convergence – Foreign Ideas, Foreign Influences, and English Law on the Eve of the 21st Century*, (Oxford, 1994). This gradual convergence, affecting all expressions of European legal systems, is equally apparent in the criminal justice field.

Criminal procedure is in fact the last resort against a potentially barbaric mechanism of state repression. The ultimate question then arises: can a civilized society ever accommodate the barbarization of a naturally humanistic process; of that particular process that draws the line between democratic societies and totalitarian regimes? Those advocating the legalization of torture might have to answer this question first.

Index